INVERTEBRATES - EXPERIMENTAL MODELS IN TOXICITY SCREENING

Edited by **Marcelo L. Larramendy**
and **Sonia Soloneski**

Invertebrates - Experimental Models in Toxicity Screening

http://dx.doi.org/10.5772/60790
Edited by Marcelo L. Larramendy and Sonia Soloneski

Contributors

Sajal Ray, Paulo Roger Lopes Alves, Elke Jurandy Bran Nogueira Cardoso, Arun Solayan, Chavon Walters, Edmund Pool, Vernon Somerset, Roberto Rico-Martinez, Mario Alberto Arzate-Cárdenas, Daniel Robles-Vargas, Ignacio Alejandro Pérez-Legaspi, Jesús Alvarado-Flores, Gustavo Emilio Santos-Medrano

CBS, Edition **2017**

Published by InTech

Janeza Trdine 9, 51000 Rijeka, Croatia

Publishing Process Manager Iva Lipovic

Technical Editor InTech DTP team

Cover Designer InTech Design Team

Additional hard copies can be obtained from orders@intechopen.com

Invertebrates - Experimental Models in Toxicity Screening,
Edited by Marcelo L. Larramendy and Sonia Soloneski
p. cm.
ISBN 978-953-51-2246-3

INTECH
open science | open minds

Contents

Preface

A wide range of chemicals present in day-to-day life, including petroleum products, petrochemicals, surfactants, pesticides, pharmaceuticals, medicines, household products, nanomaterials, food additives, agricultural run-off, by-products of farming, and industrial wastes, among others, are released into the environment in large quantities. However, the available information about their toxic effects on biological components of the environment is inconclusive.

These compounds can be hazardous if not used appropriately, and many of them (if not all) may present a real risk to the environment of contaminating soil, water, and air. Furthermore, by increasing their jeopardizing effects, anthropogenic activities are daily introducing extensive amounts of these compounds into the environment. Most of the pollutants in the different environmental compartments exert their effects through cytotoxic, genotoxic, and metabolic toxic mechanisms. Accordingly, current awareness of the real and potential hazards of pollutants in the environment has a high interest in the use of invertebrate and vertebrate species as indicators for monitoring pollutant-induced deleterious environmental effects. Additionally, it is known that most of the environmental pollutants not only affect target organisms but also concomitantly exert negative effects on nontarget species.

In pollution studies, there is an increasing interest in biomonitoring markers to provide measurements of biological exposure to pollutants. To achieve this goal, several end points for testing toxicity, cytotoxicity, and genotoxicity have been used in aquatic and terrestrial invertebrates to assess the impact of pollution on contaminated areas (in situ assays) and to screen for xenobiotics after direct or indirect exposure (*in vivo* assays).

In this context, invertebrates have been used for decades in acute and chronic toxicity tests for hazard identification. They can be very efficient screening systems and have a major role in toxicity research because certain aspects of their biology, physiology, and genetic characteristics are sufficiently similar to those of vertebrates.

This book, *Invertebrates - Experimental Models in Toxicity Screening* , is intended to provide an overview of the use of conventional and nonconventional invertebrate species as experimental models for the study of different toxicological aspects induced by environmental pollutants in both aquatic and terrestrial ecosystems. We aimed to compress information from a diversity of sources into a single volume.

The chapters in this book include details of various environmental pollutant–related topics about ecological risk assessment after xenobiotic exposure. This book presents three comprehensive review chapters related to the use of aquatic and terrestrial invertebrate models for toxicity screening. The first of these chapters focuses on the use of aquatic invertebrates as a suitable model for the evaluation of the reactivity and responses of invertebrates toward xenobiotics, highlighting that the responses are species specific and related to the chemistry of the toxin. The second chapter aims to validate the use of aquatic organisms as biotic matrices for the environmental health risk evaluation of the large-scale production of nanomaterials. The third chapter summarizes the use of soil fauna species for the screening of soil pollutants. Finally, two additional chapters focusing on the contribution of rotifers as attractive organisms for ecotoxicological studies and coelenterates for the biomonitoring of coral bleaching induced by oxidative stress, respectively, are also included.

Several researchers have contributed to the publication of this book, which is of high importance to researchers, scientists, engineers, and graduate students who make use of these different investigations to understand the importance of the use of conventional and nonconventional aquatic and soil invertebrate species in environmental risk assessment. Furthermore, it is hoped that the information in the present book will be of value to those directly engaged in the handling and use of environmental pollutants and that this book will continue to meet the expectations and needs of all interested in the different aspects of toxicity screening.

Prof. Dr. Marcelo Larramendy
School of Natural Sciences and Museum
National University of La Plata
La Plata, Argentina

Dr. Sonia Soloneski
National University of La Plata
Argentina

Chapter 1

Levels of Toxicity Screening of Environmental Chemicals Using Aquatic Invertebrates - A Review

Sajal Ray

Additional information is available at the end of the chapter

http://dx.doi.org/10.5772/61746

Abstract

Effective screening of the toxicity of chemicals using living organisms has been considered as a major issue of environmental biomonitoring. The principle of toxicity screening involves the quantitation of toxin-induced shift of biological response or tissue morphology of test species both *in vivo* and *in vitro*. Most of the toxin appears to function as biological response modifiers at a defined concentration and span of exposure. In recent years, invertebrates have been gaining a special scientific attention for being utilized as suitable model for toxicity screening. Invertebrates like crab, mollusks, sponge, and earthworm have already been established as model organisms for toxicity screening and analyses. A number of environmental toxins like arsenic, pyrethroid, pesticides, heavy metals, and washing soda can be screened for their toxicities using invertebrate species. Cellular and subcellular parameters like blood cell density, lysosomal membrane stability, cellular damage, apoptosis, micronucleation, and cytotoxic response of invertebrates had been established as biomarkers of environmental toxicity. Toxin-induced histopathological and behavioral shift had been suggested as effective parameters of toxicity screening in model invertebrates. However, reactivity and responses of invertebrates toward xenobiotics are often recorded to be species specific and related to the chemistry of the toxin. Current article reviews different levels of toxicity screening using invertebrate as test model.

Keywords: Sponge, mollusks, crab, arsenic, washing soda, pesticides, hemocytes

1. Introduction

Current global environment is characterized by the presence of diverse chemical compounds of inorganic and organic in nature [1]. Multiple human activities have been identified as the sources of origin of these environmental xenobiotics. Various industrial processes and agricultural activities result in the rapid and precarious contamination of the terrestrial and aquatic ecosystems. Environmental toxins after generation from sources of origin enter the specific ecological compartments following the nonspecific and specific patterns of dispersion [2, 3]. Environmental dispersion of xenobiotics largely depend on their physicochemical characteristics which largely influence the distribution and fate of the toxic compounds related to half life, polarity, environmental and cellular degradability and toxicokinetics [4]. Upon entering the cellular environment, many of these environmental chemicals undergo enzyme guided xenometabolic transformation chemically termed as biotransformation. Xenometabolism of environmental chemicals may exhibit dual modes of transformation i.e. bioactivation and bioinactivation. Both bioactivated and bioinactivated environmental metabolites are the enzymatic products of biochemical degradation. Principal objective of xenometabolism involves transformation of nonpolar chemical compounds into a product with higher polarity for facilitation of urinary excretion. However, classification of environmental toxins distributed in the current global environment had never been an easy task for the environmental chemists. Rapid and uninterrupted contaminations of the global ecosystem by newer generations of chemical compounds have been identified as a serious challenge and require effective screening of their toxicity in suitable animal models. Major classes of environmental toxins that have been characterized by the chemists and biologists include acids, alkalis, pesticides [1], nuclear fall outs, diverse organic compounds etc. With the rapid shift in the overall physicochemical characteristics of global environment, a continuous search for suitable biological systems for toxicity screening of xenobiotics is being carried out by the toxicologists employing invertebrate species. Scientific information available in the recent years indicates the suitability of invertebrates as "test organism" for toxic screening of environmental chemicals (Figure 1).

2. Environmental chemicals

Current global environment is characterized by the presence of diverse chemical compounds of inorganic and organic in nature [1]. Multiple human activities have been identified as the sources of origin of these environmental xenobiotics. Various industrial processes and agricultural activities result in the rapid and precarious contamination of the terrestrial and aquatic ecosystems. Environmental toxins after generation from sources of origin enter the specific ecological compartments following the nonspecific and specific patterns of dispersion [2, 3]. Environmental dispersion of xenobiotics largely depend on their physicochemical characteristics, which largely influence the distribution and fate of the toxic compounds related to half life, polarity, environmental, and cellular degradability and toxicokinetics [4]. Upon entering the cellular environment, many of these environmental chemicals undergo enzyme-guided xenometabolic transformation chemically termed as biotransformation. The xenometabolism of environmental chemicals may exhibit dual modes of transformation, i.e., bioactivation and bioinactivation. Both bioactivated and bioinactivated environmental metabolites are the enzymatic products of biochemical degradation. The principal objective of xenometabolism involves the transformation of nonpolar chemical compounds into a product with higher polarity for facilitation of urinary excretion. However, the classification of

http://dx.doi.org/10.5772/61746

environmental toxins distributed in the current global environment had never been an easy task for the environmental chemists. Rapid and uninterrupted contaminations of the global ecosystem by newer generations of chemical compounds have been identified as a serious challenge and require effective screening of their toxicity in suitable animal models. Major classes of environmental toxins that have been characterized by the chemists and biologists include acids, alkalis, pesticides [1], nuclear fall outs, diverse organic compounds, etc. With the rapid shift in the overall physicochemical characteristics of global environment, a continuous search for suitable biological systems for toxicity screening of xenobiotics is being carried out by the toxicologists employing invertebrate species. Scientific information available in the recent years indicates the suitability of invertebrates as "test organism" for toxic screening of environmental chemicals (Figure 1).

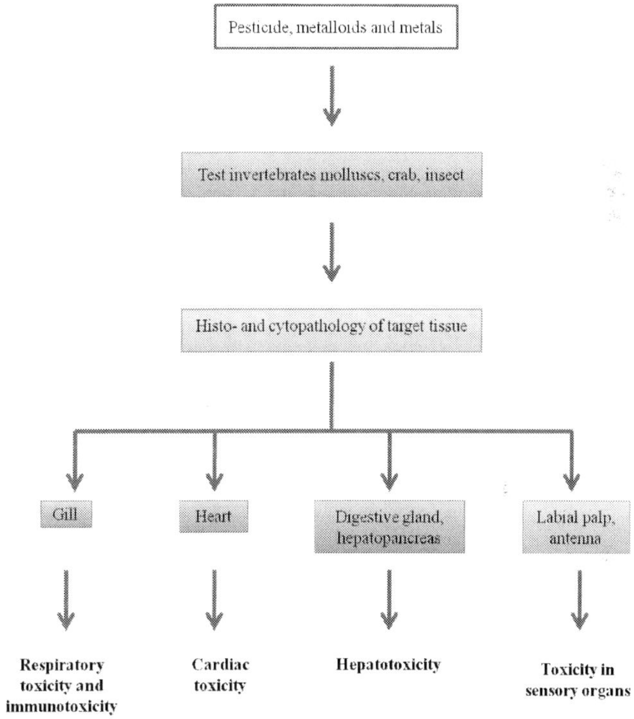

Figure 1. Cellular and organ pathology as major screening parameters of toxins.

2.1. Invertebrates and toxicity screening

The anatomical variation of body cavity or coelom greatly influences the physiological and adaptational attributes of invertebrates. Invertebrates without coelomic cavity are termed as acoelomate, whereas the species with a true coelom is termed as eucoelomate. Pseudocoelomate is a characteristic intermediate group without a true coelom. Invertebrates with open mode of circulatory system evolved blood cells termed as hemocytes and coelomocytes. Circulating blood cells of molluscan hemocoel is known as hemocytes, whereas coelomocytes (Figure 2) are the coelomic cells recorded in the annelids. Recent scientific reports indicate the efficacy of hemocytes and coelomocytes of invertebrates as ideal tools for screening the toxicity of environmental chemicals of known or less known toxicity. Many of these screening methodologies involved both *in vitro* and *in vivo* modes of laboratory testing.

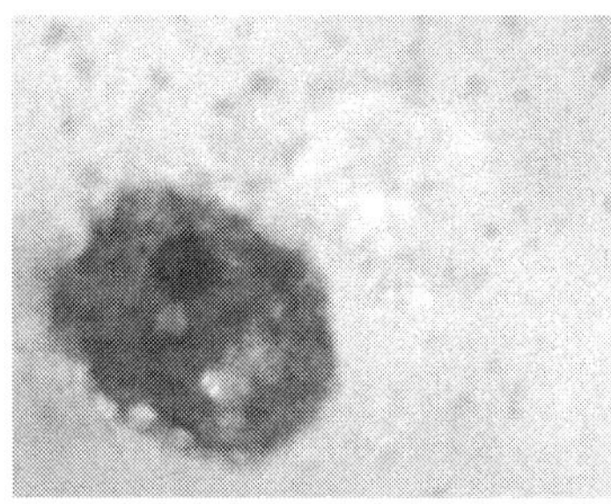

Figure 2. Morphofunctional attributes of earthworm coelomocyte reflects the efficacy of soil invertebrates to act as model organisms to screen the toxicity of agrochemicals.

2.1.1. Morphological aberrations of invertebrate blood cells as a measure of the toxicity of environmental chemicals

Hemocytes or coelomocytes are the main types of circulatory cells of invertebrates. They perform diverse types of physiological functions and include transport and carriage of gases, nutrients and bioactive substances, nonself recognition, and deactivation of environmental pathogens and toxins. Discrete subpopulations of invertebrate blood cells act as immunocytes and are involved in elicitation of immunological reactivity against environmental toxins and pathogens. Toxins with diverse chemical identities are capable of generating varied degrees of morphological damage or alteration of blood cells of invertebrates [2, 5, 6, 7]. The microscopic examination of morphological aberrations of blood cells thus serves as a unique tool for the testing of the toxicity of environmental chemicals both *in vivo* and *in vitro*. The treatment of the test organisms with measured quantity of chemical toxins *in vivo* provides satisfactory result in experimental toxicology for toxicity testing [2]. The testing of toxicity employing short- or long-term cell or tissue culture technology (Figure 3) occasionally provides an opportunity for the toxicologists in screening the toxicity of xenobiotics. However, invertebrate cell and tissue culture technology appears to be a challenging domain due to the lack of the scientific information and species-specific cellular response in artificial culture media. Hemocytes of aquatic mollusks upon treatment with xenobiotics yield multiple morphological aberrations like vacuolation, membrane disintegration, cellular disruption, and shift in size and shape [2].

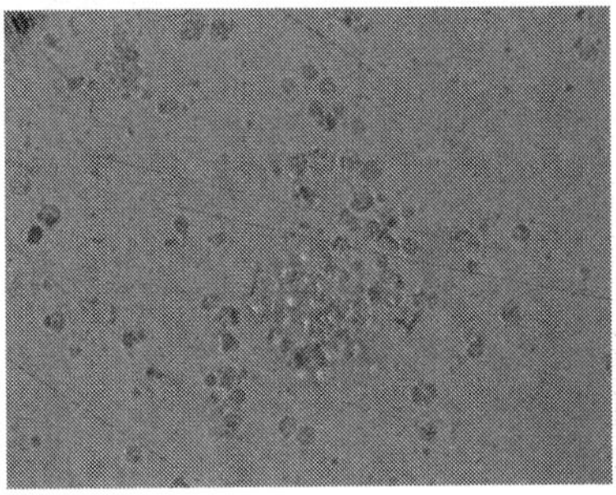

Figure 3. Molluscan hemocyte culture is an effective technology of toxicity screening *in vitro*.

2.1.2. Functional parameters

Defined functional parameters of hemocytes, coelomocytes, or other body cells may also serve as a tool for screening the toxicity of chemical compounds. Studies of the functional attributes of invertebrate cells require advanced level of bioinstrumentation. Some of the established functional parameters of invertebrate cells include phagocytic response, aggregation response, cytotoxicity, cell doubling time, lysosomal membrane fragility, etc. Phagocytosis (Figure 4) is established as an innate immunological response reported in all major invertebrate Phyla [8, 9]. It is characterized by sequential-like recognition, chemotaxis, contact, and internalization followed by the degradation of the target in the cell's interior. The degree of phagocytic response can be quantitated by suitable cellular index. The toxicity of multiple environmental chemical can suitably be estimated by determining the phagocytic index of invertebrate immunocytes against ideal control. Cellular aggregation (Figure 5) is also recognized as a metabolic behavior of cells against chemicals or foreign particulates of toxic nature. Various molluscan test species exhibit the aggregation response of hemocytes upon the exposure of environmental pathogens and toxins [10].

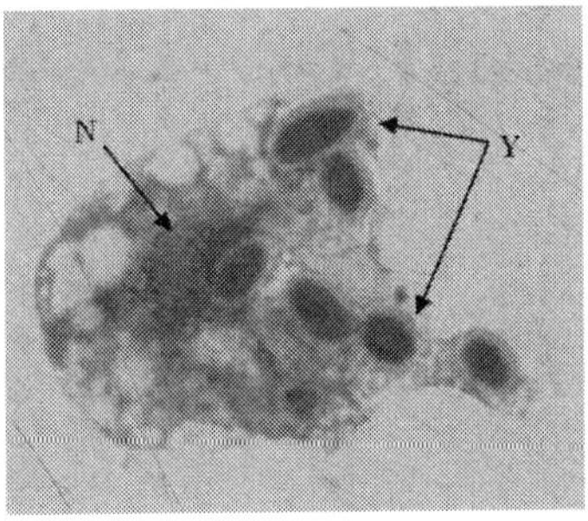

Figure 4. Phagocytic response of sponge cells is an established tool of toxicity screening. N—sponge cell nucleus, Y—yeast particle.

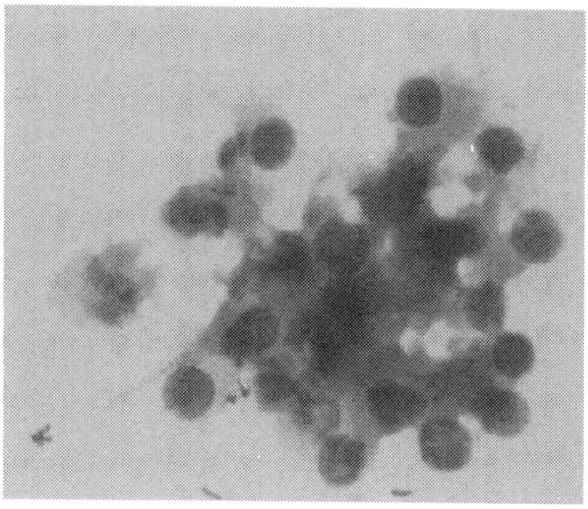

Figure 5. Aggregation response of molluscan hemocytes is often modulated by chemical toxins.

2.1.3. Dynamics of blood cell density as a measure of toxic response

Blood cell homeostasis of invertebrates is a function of multiple cellular processes. These include rate of hematopoiesis, mitosis, cell migration, and aggregation tendency of hemocytes or coelomocytes. The maintenance of steady state of total count of blood cell depends on the overall health and immunological status [11] of the model organism. The morphological variation of blood cells is not uncommon in the phylogeny. Therefore, differential cell count may also act as a suitable parameter of toxicity screening. However, a discrete scheme of the morphological classification of hemocyte and coelomocyte appears to be absent in many invertebrates [8]. In many cases, the lack of proper nomenclature of hemocyte subpopulation creates a scientific problem in the identification of cells needed for toxicity screening. However, in invertebrates, hemocyte populations like blast cells had been identified as a suitable candidate for the testing of toxicity both *in vivo* and *in vitro* at cellular level.

2.1.4. Nuclear aberration and lysosomal membrane fragility of hemocytes and sponge cells

The natural habitats of aquatic invertebrates are often contaminated with diverse toxins of known or less known chemistry. Thus, the hemocytes are being continuously exposed to environmental toxins of varied concentrations. Such a situation often leads to onset of genotoxicity in many of the species belonging to Crustacea and Mollusca. Hemocytes, upon exposure to environmental toxins, may present multiple nuclear aberrations like micronucleation, binucleation (Figure 6), or trinucleation; karyolysis; chromatin condensation; pycnosis; etc. The degree of toxin exposure is often correlated with the magnitude or frequency of nuclear aberrations *in vivo* [5]. Environmental toxins like arsenic, pyrethroid, and alkaline chemical compounds may generate such types of nuclear anomalies in invertebrate. Following this principle, scientists proposed nuclear aberration as a suitable tool for the testing of the toxicity of environmental chemicals. The toxin-induced fragility of lysosomal membrane of hemocytes is in report [2, 7]. Toxin exposure often leads to damage of lysosomal membrane leading to the release of hydrolases in the cytoplasm. This may result in impairment in the structural profile of cells and tissues. The degree of lysosomal membrane fragility can be quantitated by neutral red retention assay [2, 5, 7] in invertebrates. The lysosomal membrane stability of molluscan hemocytes and sponge cells has been claimed as biomarkers of environmental toxicity (Figure 7).

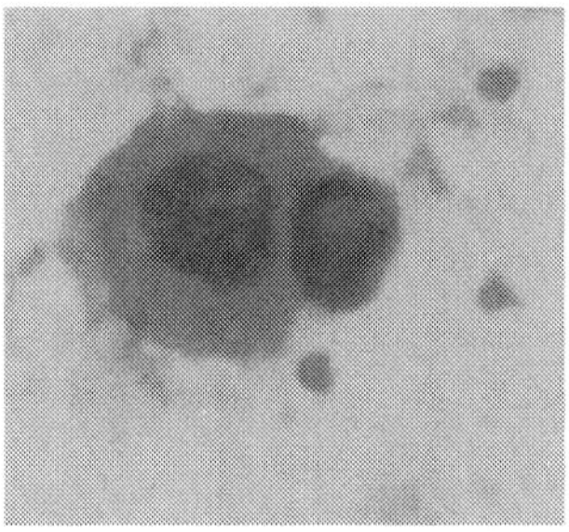

Figure 6. Pesticide-induced binucleation in molluscan hemocytes is a marker of screening genotoxicity.

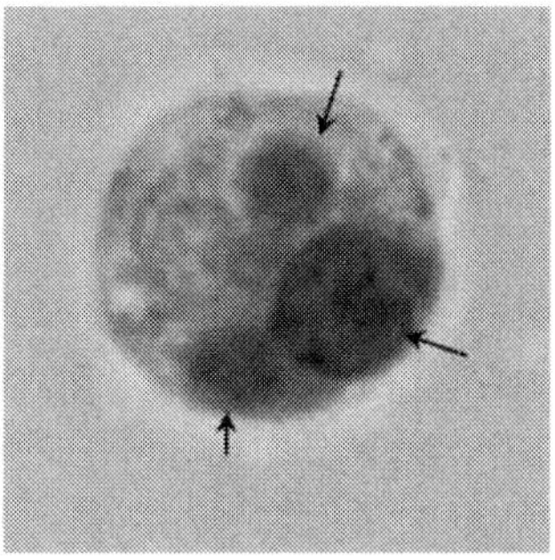

Figure 7. Lysosomal membrane stability of cells is an established parameter of screening ecotoxicity in sponge [7]. Arrows indicate the diffused neutral red probe in cytoplasm of sponge cell.

2.1.5. Target enzymes

The activity of enzymes as biocatalysts in invertebrate tissues or cells may be considered as effective parameters of assessment of the toxicity of chemicals. Target enzymes of toxins are reported to be located in the cellular or extracellular compartments. Therefore, prior to a biochemical estimation of the activity of these enzymes, physicochemical characterization, and lysate preparation methodology appear to be an important step for the toxicologists. Principal enzymes that are considered and established for toxicity screening include acetylcholinesterase [12], glutamate oxaloacetate transaminase, glutamic pyruvic transaminase [13], ATPase, phosphatases [13, 7], etc. The activity of acetylcholinesterase is established as an effective parameter of examining the neurotoxicity of many pesticides and allied compounds. The activity of enzymes like glutamate oxaloacetate transaminase, glutamic pyruvic transaminase, ATPase, and phosphatase appeared to be effective and sensitive toward the exposure of many metabolic toxins.

2.1.6. Histopathology

Selected organs or tissues of aquatic invertebrates undergo histological alterations under chronic, subchronic, or acute toxin exposure. Pathological changes of target tissues often provide an excellent scope of assessment of the nature and exposure of chemical toxins (Figure 8). Histopathological analyses had been established as a useful method of toxicity screening in both invertebrates and vertebrates. Histopathology also provides an early signal of pathogenesis and environmental toxicity of diverse toxins of known or unknown chemistry in invertebrates. The magnitude of toxicity by xenobiotics depends on chemical characteristics, route of entry, dosage, span of exposure, and toxicokinetics.

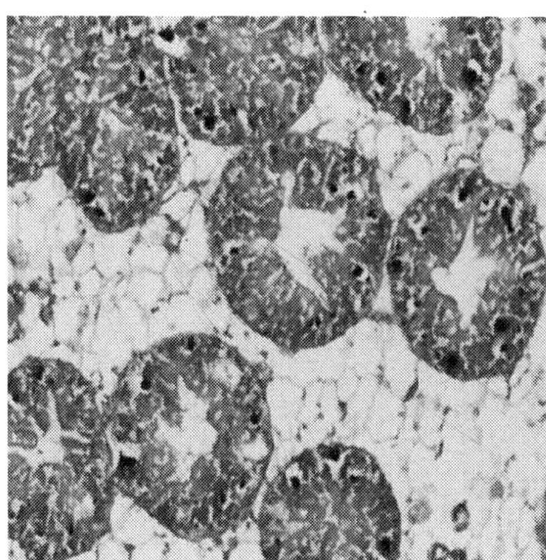

Figure 8. Histopathological analyses of digestive gland of marine mollusk, *Telescopium telescopium*, is an effective screening parameter of aquatic diesel toxicity.

Many aquatic invertebrates respire through gill. Gills are highly vascularized membranous structures involved in important physiological process like gaseous exchange with environment, filter feeding, and immunosurveillance [13]. Gills, being relatively exposed to the external environment, interact intimately with the toxic compounds dissolved or suspended in water. This characteristic feature permits gills to act as an ideal organ for histopathological analyses under toxin exposure. In many species belonging to Mollusca and Crustacea, the histopathology of gill had been reported as a suitable procedure of screening the toxicity of chemical pollutants. Toxin-induced histopathological damage of gill is an indicator in the impairment of the associated functional status of the test organisms. The exposure of the gills of the aquatic bivalves and gastropods to environmental arsenic [13], detergent, pyrethroids, and azadirachtin-based pesticides often lead to the appearance of the hyperchromatic anaplastic cells, clogging of water channels, and lamellar membrane disruption in varied degree.

The exposure of toxins like arsenic and cypermethrin yielded substantial histopathological changes in the heart of the aquatic mollusks [12]. Pathological changes in the auricle and ventricle were prominent under the acute or semiacute treatment of toxins. The unrestricted exposure of test organisms to toxin may thus lead to onset of cardiac toxicity in invertebrates. Histopathological analyses in association of functional assessment of target organs provide an excellent premise of screening the toxicity of xenobiotics. Apart from vital organs like gill and heart, histopathological examination of organs like labial palp [14] of aquatic mollusks had been suggested as an effective procedure of the toxicity screening of pesticides.

2.1.7. Behavioral attribute as a measure of toxicity screening

Behavior in general is considered as a manifestation of physiological performance of organism in both natural and stressed conditions. Invertebrate ethology is relatively a less studied area with limited scientific information. However, behavioral response exhibited by a test species under toxic exposure has been indicated as a method of screening environmental toxicity [14]. Significant deviation in the normal ethogram due to toxin exposure may thus be considered as an effective method of assessment of toxin-induced stress in the test invertebrate.

In mollusks, salient behaviors like relative mobility, aggregation, and mucus release had been established as screening parameters of toxicities of azadirachtin, a neem-based biopesticide. The relative mobility of an organism is often associated with many biological functions like food gathering, mate approach, predatory escape response, etc. [15], in invertebrates. The exposure of toxin adversely affects these parameters which interfere with the functional performance of them and their natural habitat. "Grouping" or aggregation and mucus secretion response had been examined in bivalve and gastropods [16]. Screening of the behavioral toxicity of chemical compounds employing aquatic invertebrates thus appears to be a novel methodology in applied toxicology.

In recent years, invertebrates in general have been gaining a special scientific attraction as model organisms for screening environmental toxicity. Invertebrates occupy diverse habitats with multiple physicochemical characteristics. Their response to a particular ecosystem often appears to be highly specific. Considering the diverse range of global aquatic environment an estuarine invertebrate may appear to be unsuitable for bioresponse assay of toxin of freshwater ecosystem. This evolutionary and adaptational specificity renders the invertebrates for being ideal model organism for screening of a particular toxin distributed in a definite ecological area. In screening the toxicity of chemicals employing invertebrate species, toxicologists should utilize both *in vivo* and *in vitro* screening systems. Cell culture technology provides an excellent opportunity in screening the toxicity of xenobiotics *in vitro*. However, invertebrate cell and tissue culture is a difficult scientific domain due to variation in the metabolic and other attributes of invertebrate cells. The chemical sensitivity of organisms had been proposed as a primitive character expressed in most all species of major invertebrate Phyla.

Thus, invertebrate in general bears a bright prospect to serve as effective test models for screening the chemical toxicity of xenobiotics. Many of the invertebrates are less researched groups of organisms of inadequate cellular and subcellular information. An in-depth research at the levels of organ, tissue, and cell is required for the proper utilization of aquatic invertebrates as model candidates for toxicity screening.

Acknowledgements

Doctor Mitali Ray and all of my doctoral students are thankfully acknowledged for their scientific effort in establishing many toxicity screening parameters in invertebrates. Departmental funds received from DST-FIST and University Grant Commission (SAP DRS II) of the Government of India are also acknowledged.

Author details

Sajal Ray*

Address all correspondence to: raysnailmail@rediffmail.com

Aquatic Toxicology Laboratory, Department of Zoology, University of Calcutta, Kolkata, West Bengal, India

References

[1] Ray S, Ray M, Chakraborty S, Mukherjee S. Immunotoxicity of environmental chemicals in the pearl forming mussel of India-a review. In: McGevin LE (ed.) *Mussels: Anatomy, Habitat and Environmental Impact*. Nova Science Publishers Inc. NY, USA; 2011. pp. 429–440.

[2] Ray M, Bhunia AS, Bhunia NS, Ray S. Density shift, morphological damage, lysosomal fragility and apoptosis of hemocytes of Indian molluscs exposed to pyrethroid pesticides. *Fish Shellfish Immunol* 2013b; 35:499–512.

[3] Mukherjee S, Ray M, Ray S. Immunotoxicity of washing soda in a freshwater sponge of India. *Ecotoxicol Environ Saf* 2015b; 113:112–123.

[4] Ray S, Mukherjee S, Bhunia NS, Bhunia AS, Ray M. Immunotoxicological threats of pollutants in aquatic invertebrates. In: Larramendy, ML (ed.) *Emerging Pollutants in the Environment—Current and Further Implications*. InTech. 2015. pp.149–167.

[5] Chakraborty S, Ray S. Nuclear morphology and lysosomal stability of molluscan hemocyte as possible biomarkers of arsenic toxicity. *Clean* 2009; 37(10):769–775.

[6] Chakraborty S, Ray M, Ray S. Cell to organ: Physiological, immunotoxic and oxidative stress responses of *Lamellidens marginalis* to inorganic arsenite. *Ecotoxicol Environ Saf* 2013; 94:153–163.

[7] Mukherjee S, Ray M, Dutta MK, Acharya A, Mukhopadhyay SK, Ray S. Morphological alteration, lysosomal membrane fragility and apoptosis of the cells of Indian freshwater sponge exposed to washing soda (sodium carbonate). *Ecotoxicol Environ Saf* 2015c; 122:331–342.

[8] Ray M, Bhunia NS, Bhunia AS, Ray S. A comparative analyses of morphological variations, phagocytosis and generation of cytotoxic agents in flow cytometrically isolated hemocytes of Indian molluscs. *Fish Shellfish Immunol* 2013a; 34:244–253.

[9] Mukherjee S, Ray M, Ray S. Phagocytic efficiency and cytotoxic responses of Indian freshwater sponge (*Eunapius carteri*) cells isolated by density gradient centrifugation and flow cytometry: a morphofunctional analysis. *Zoology* 2015a; 118:8–18.

[10] Mukherjee S, Ray M, Ray S. Anticoagulant and carbohydrate induced interference of aggregation of mussel haemocyte under azadirachtin exposure. In: McGevin LE (ed.) *Mussels: Anatomy, Habitat and Environmental Impact.* Nova Science Publishers Inc. NY, USA; 2011. pp. 441–451.

[11] Chakraborty S, Ray M, Ray S. Sodium arsenite induced alteration of hemocyte density of *Lamellidens marginalis*—an edible mollusc from India. *Clean* 2008; 36(2):195–200.

[12] Chakraborty S, Ray M, Ray S. Arsenic toxicity: a heart breaking saga of a freshwater mollusc. *Tissue Cell* 2012; 44:151–155.

[13] Chakraborty S, Ray M, Ray S. Toxicity of sodium arsenite in the gill of an economically important mollusc of India. *Fish Shellfish Immunol* 2010; 29:136–148.

[14] Das K, Ray M, Ray S. Shift in foot protrusion frequency, cholinergic impairment in pedal ganglia and histopathology of labial palp in freshwater mussel exposed to cypermethrin. In: Gupta VK (ed.) *Perspective in Animal Ecology and Reproduction.* Daya Publishing House, New Delhi, India. 2015. pp. 123–136.

[15] Kamble SB, Kamble NA. Behavioural changes in freshwater snail *Bellamya bengalensis* due to acute toxicity of copper sulphate and *Acacia sinuate. Int J Sci Environ* 2014; 3:1090–1104.

[16] Akarte SR, Mane UH. Toxicity of Folithin to three freshwater bivalve molluscs in different seasons. *J Environ Biol* 1968; 9:27.

Chapter 2

Nanotoxicity in Aquatic Invertebrates

Chavon Walters, Edmund Pool and Vernon Somerset

Additional information is available at the end of the chapter

http://dx.doi.org/10.5772/61715

Abstract

Due to their unique properties, nanomaterials (NMs) are being incorporated in several applications including consumer products, electronics, pesticides and the pharmaceutical industry. As such, the rapid development and large-scale production of NMs has inspired concerns regarding their environmental health risks. In order to address these concerns, there has been a rapid development in the methods of toxicity testing of NMs, specifically in aquatic organisms. Understanding the unique properties of nanoscale materials has proven to be a particular important aspect of their toxicity. Properties such as surface area, surface coating, surface charge, particle reactivity, aggregation and dissolution may affect cellular uptake, *in vivo* reactivity and distribution across tissues. The behaviour of NPs is influenced by both the inherent properties of the NP as well as environmental properties (such as temperature, pH, ionic strength, salinity, organic matter). As such, this chapter describes methodologies of NM characterization in exposure media and NM *in vivo* toxicity experimental procedures under variable environmental conditions (with special emphasis on temperature).

Keywords: Toxicity, Nanotoxicity, Nanomaterials, Nanoparticles, Silver nanoparticles, Temperature, Aquatic invertebrates

1. Introduction

There has been extensive growth in nanoscale technology in the last few decades, to such a degree that nanomaterials (NMs) have become a constituent in a wide range of manufactured commercial and domestic products. This surge has resulted in uncertainties regarding their environmental impact due to the significant increases in the amount of NMs released into the environment [1] through intentional and unintentional releases. Like many other toxins, the aquatic environment is particularly vulnerable as it acts as a sink for nanoparticles (NPs) [2]. The escalating growth of NMs has not advanced without efforts to understand its properties. Despite the dramatic advances in both the production and application of NMs, very little is known regarding their interaction with and effects on environmental and human health. Given the lack in scientific knowledge, particularly under various environmental conditions, it is often difficult to accurately assess the potential exposure pathways to ecological receptors.

Silver nanoparticles (AgNPs) are the most widely used metal NPs, present in several consumer products largely due to their antibacterial properties. It is estimated that the annual production exceeds 1000 tons/year [3]. The increased use of AgNPs in consumer products (e.g. textiles, cosmetics and personal hygiene), household appliances (e.g. washing machines and vacuum cleaners) and medical equipment have led to their increased release into the environment, thereby posing an environmental risk and human health concern.

When AgNP is discarded, it can enter the environment as aggregates and soluble ions, which can be highly toxic to aquatic organisms. The dissolution of AgNPs is a significant process determining AgNPs effects in the aquatic environment and its organisms. Although environmental concentrations of AgNPs have not been determined, it is estimated that more than 15% of Ag released into waters will come from plastics and textiles containing AgNPs [4]. In addition, it is predicted that concentrations of AgNPs in natural waters range from 0.03 to 500 ng/L [5]. A fundamental question is whether AgNPs remains in the particle phase in the environment following dissolution or whether it poses an additional risk.

Silver NPs are known to induce the production of reactive oxygen species (ROS) [6–8]. Also, since AgNPs are oxidized to ionic Ag (Ag^+), it is still unclear whether the effects of ROS can be attributed to Ag^+ release or to the AgNP itself [9, 10]. To cope with these and other stressors, aquatic organisms are able to modulate their physiological and biochemical metabolism through antioxidant defences, which consist mainly of antioxidant enzymes that reduce the damaging effects of hydrogen peroxide (H_2O_2) to prevent the production of hydroxyl radicals ($HO\cdot$), the most damaging oxygen species [11]. These oxidative stress biomarkers have been widely used as "early warning" signs of environmental stress.

Assessing the ecotoxicity of NMs is a challenging task. Inexpensive, rapid and reproducible methods are preferred, and a coordinated standardization could help in avoiding the waste of resources. Standardized tests established by the Organization for Economic Co-operation and Development (OECD) and the U.S. Environmental Protection Agency (U.S. EPA) have protocols for testing aquatic, terrestrial and microbial organisms. However, these tests were established considering conventional chemicals and not NMs. The general consensus in the scientific community is that the basis of these standardized tests (i.e. test organism, endpoints) may generally work for NMs but would require some modifications. Nevertheless, NMs remain very poorly tested in contrast with their larger counterparts; the main difficulties in assessing toxicity are due to their colloidal nature and unique properties. The behaviour of NPs is collectively influenced by inherent (NP size, shape, surface area, surface chard, crystal structure, coating, solubility/dissolution) and environmental factors (temperature, pH, ionic strength, salinity, organic matter).

The potential implications and effects of nanotechnology and NMs on environmental and human health is an important issue of global concern. The focus of the proposed research is to investigate the effects of AgNPs when exposed to simulated climate changes (such as extreme temperatures), thus mimicking the conditions experienced naturally in the environment during potentially extreme conditions. The research areas which this research aims to address include NP fate and transport, bioavailability and ecotoxicology (or nanoecotoxicology).

1.1. Oxidative stress

Oxidative stress is defined as an imbalance between the production of reactive oxygen species (ROS) and the cells' ability to reduce ROS, which may be a result of increased ROS production, a decrease in the cell's defence mechanisms or a combination of both. Disturbances in the normal redox state of cells may cause toxic effects through the production of peroxides and free radicals that in turn damage cells, including proteins, lipids, and DNA. Because certain reactive oxidative species act as cellular messengers in redox signalling, oxidative stress may lead to disruptions in normal mechanisms of cellular signalling. ROS refers to oxygen free radicals, partially reduced intermediates of the four electron reduction of oxygen to water, i.e. superoxide anions ($\cdot O_2^-$), hydroxyl radicals ($\cdot OH$) and the nonradical active species hydrogen peroxide (H_2O_2). Aerobic organisms, which derive their energy from the reduction of oxygen, are particularly susceptible to the damaging actions of the small quantities of $\cdot O_2^-$, $\cdot OH$ and H_2O_2 that form during the metabolism of oxygen [12].

Biomarkers of oxidative stress can offer an early warning sign for exposure to xenobiotics. Biomarkers such as enzyme activity are widely used for environmental monitoring. Measurements in this category range from markers related to redox status (e.g. superoxide dismutase (SOD activity), reproduction-associated proteins (e.g. vitellogenin) and stress response pathways (e.g. antioxidant responses and heat shock protein) [13]. Figure 1 represents a schematic of the major oxidative pathways. A brief description of the oxidative stress biomarkers used in this study follows in Section 1.2.

1.2. Antioxidant defence system

A number of defence mechanisms have evolved to provide a balance between production and removal of ROS. Cells have a variety of elaborate defence mechanisms to restore the harmful effects of ROS. The removal of foreign substances (xenobiotics) from cells is catalyzed by several enzymes, particularly Phase I and Phase II enzymes. Phase I enzymes initiate the detoxification process by chemically transforming lipid soluble compounds into water soluble compounds in preparation for Phase II detoxification [14] (Equation 1). These include the cytochrome P450 (CYP450) enzymes which are responsible for most Phase I reactions. CYP450 are typically found in the membranes of the endoplasmic reticulum (microsomes) within liver cells (hepatocytes). Activity of Phase I enzymes can typically lead to an increase in ROS production. Antioxidant enzymes facilitate the removal of these resulting ROS molecules and reactive chemical intermediates. The action of CYP enzymes results in the production of O_2, which consequently can be metabolized by SOD to hydrogen peroxide (H_2O_2) and oxygen (O_2) (Equation 2), which can in turn be reduced to water (H_2O) and O_2 by CAT (Equation 3) or glutathione peroxidase (GPx) (Equation 4) [15].

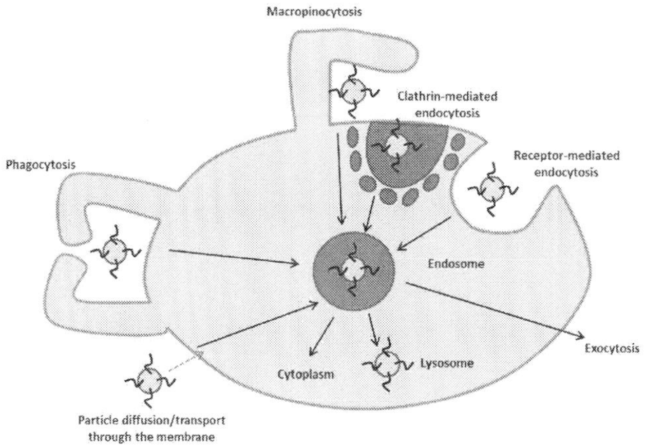

Figure 1. Schematic diagram of oxidative stress (adapted from www.sigmaaldrich.com).

$$H + O_2 + 2e^- + 2H^+ \rightarrow OH + H_2O \quad (1)$$

$$2O_2^- + 2H^+ \rightarrow 2H_2O + O_3 \quad (2)$$

$$2H_2O_2 \rightarrow 2H_2O + O_2 \quad (3)$$

$$2GSH + H_2O_2 \rightarrow GSSG + 2H_2O \quad (4)$$

Nonenzymatic antioxidants also play a role in detoxification. The tripeptide glutathione exists as reduced glutathione (GSH) and oxidized glutathione (GSSG). Reduced glutathione (GSH) is a major tissue antioxidant that provides reducing equivalents for the GPx catalyzed reduction of lipid hydroperoxides to their corresponding alcohols and H_2O_2 to H_2O. When cells are exposed to increased levels of oxidative stress, GSSG accumulates and the GSSH:GSH ratio increases. This increased ratio of GSSG-to-GSH is indicative of oxidative stress. The reaction catalyzed by glutathione peroxidase requires GSH as a substrate and is determined by the ratio of GSSG:GSH. This ratio is an indication of the redox state of cells [16] and is important to ROS detoxification.

1.3. Uptake and accumulation of silver nanoparticles

Once introduced into aquatic ecosystems, the fate and transport of AgNPs and its uptake by aquatic biota depends on several factors. NP properties (such as size, shape and coatings), and

http://dx.doi.org/10.5772/61715

water chemistry (such as dissolved organic carbon, ionic strength, pH, temperature) will largely influence the extent to which these particles will either remain in suspension, partition to dissolved organic carbon in the water column, form aggregates or adsorb to suspended particles [17].

In aquatic organisms, the major routes of entry are via ingestion or direct passage across the gills and other external surface epithelia. Recent studies with *Daphnia magna* have indicated that AgNPs may be internalised by these routes [18]. At the cellular level, internalisation of NP occurs via endocytosis. Mechanisms of cellular uptake of NPs are described in Figure 2. Three main mechanisms are responsible for NP uptake: phagocytosis, macropinocytosis and receptor-mediated endocytosis [19]. During phagocytosis (a specific form of endocytosis), particles are taken up the invagination of the plasma membrane. Jayaseelan et al. [20] showed internalization of nickel NPs in Mozambique tilapia (*Oreochromis mossambicus*), demonstrating the feasibility of uptake via this route. Macropinocytosis involves the internalization of a larger area of membrane. Other forms of endocytosis include clathrin- and receptor-mediated endocytosis. Nanoparticles can also enter cells by diffusion or transport through the cell membrane, resulting in particles located freely in the cytoplasm [21].

The accumulation of NPs by aquatic organisms is dependent on both the uptake and the elimination (detoxification) of the NP out of the organism [22]. Processes which regulate the bioaccumulation (and bioavailability) of AgNPs include: the concentration of the AgNP, the physicochemical properties the AgNP, the characteristics of the environment such as abiotic factors, the route of exposure, the biology and functional ecology of the organism involved and exposure duration [23].

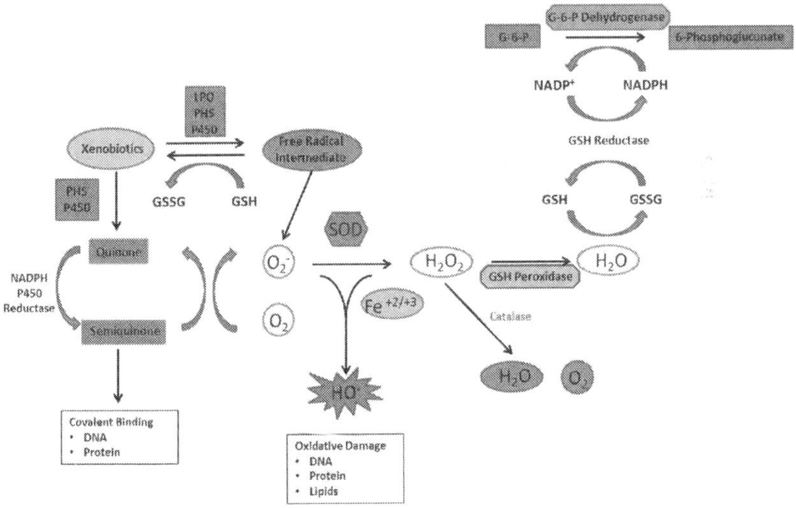

Figure 2. Mechanisms of cellular uptake of NPs.

2. Crustaceans and exposure routes in ecotoxicology

Potamonautes is a genus of freshwater crabs in the family Potamonautidae. They represent a fairly common species, having the widest distribution in sub-Saharan Africa, and are widespread under boulders in the middle and lower reaches of rivers and other freshwater water courses. As with other crustaceans, they have a segmented body with a rigid exoskeleton and jointed limbs, and an open vascular system in which numerous haemocytes freely circulate in haemolymph. The colour of the Cape River crab *Potamonautes perlatus* can vary from dark brown to mottled green. Freshwater crabs typically have nine pairs of gills, which lie in the two branchial chambers of the carapace [24]. The digestive system is basically composed of a foregut, midgut and hindgut. The foregut is comprised of a mouth, oesophagus and stomach, while the midgut is composed of an anterior and posterior caecum and midgut gland (hepatopancreas). The hindgut is a simple straight tube, which finishes at the anus. The reproductive system is very simple, consisting of paired gonads that open onto the ventral surface of the trunk [24].

Crustaceans show a high sensitivity to environmental stressors [25] and are therefore found to be useful bioindicators for monitoring the pollution state in aquatic environments [26]. Contaminated ecosystems induce deleterious effects on aquatic organisms. In crustaceans, the exposure routes are mainly via ingestion and adsorption to surface epithelia such as the gills. As an example, de Freitas Rebelo et al. [27] reported histopathological effects in the gills (disruption of pilaster cells and collapse of gill lamellae) of the estuarine crab *Chasmagnathus granulate* following exposure to ammonia. In addition, NPs that are taken up via ingestion through the digestive tract may accumulate in the hepatopancreas [28]. Nanoparticles may also diffuse into the circulating haemolymph and settle in a target organ.

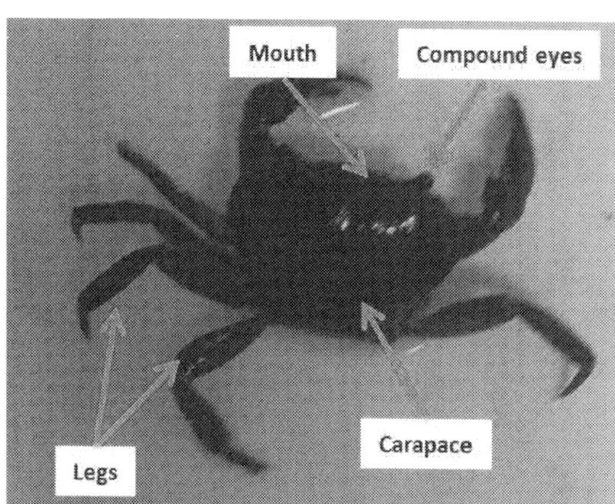

Figure 3. Dorsal view of *P. perlatus*.

3. Methodology

3.1. Characterization of nanomaterials in solution prior to *in vitro* exposure

3.1.1. Stock suspension and media preparation

AgNPs (Cat. No. 7440-22-4) were purchased from Sigma-Aldrich (Sigma Aldrich, South Africa). The product information indicated a particle size of <100 nm diameter, purity of 99.5%, a specific surface area of 5 m^2/g and density of 10.5 g/cm^3. The stock AgNP suspension (1 mg/mL) was prepared by dispersing AgNPs (1 mg dry AgNPs powder) in deionized water (1 mL) and sonicating for 5 min in an ultrasonic bath in order to disrupt any possible aggregates. The 1 mg/mL AgNP stock suspension was added to each of the plastic aquaria to obtain the appropriate final concentration.

3.1.2. SEM of nanoparticles

Scanning electron microscopy (SEM) with energy-dispersive X-ray (EDX) spectrometry capabilities was used to characterize particle size and confirm the presence of Ag. Sample preparation for SEM analysis included sonicating a dilute suspension of the AgNP in ethanol for 1 hour and dropping a portion onto a carbon surface of a SEM stub with a pipette. The droplet was allowed to dry and the SEM analysis was performed on an EVO® MA15. Samples were identified with secondary electrons and/or secondary electron images, and compositions were quantified by EDX analysis using an Oxford Instruments® X-Max 20 mm^2 detector and Oxford INCA software.

3.1.3. TEM of nanoparticles

Transmission electron microscopy (TEM) characterization was performed to obtain the NP size and morphology on a JEOL 1200-EX II electron microscope at an accelerating voltage of 120 kV. Samples were imaged with a MegaView Camera with Gatan Microscopic software with a resolution of 1376 × 1032, and two seconds exposure time. A suspension of AgNP was dissolved in ethanol, and subsequently deposited onto copper grids and air-dried. ImageJ software was used to generate a particle size distribution based on the TEM images.

3.1.4. Powder X-Ray Diffraction (PXRD)

A Panalytical X'pert Pro diffractometer was used to collect a PXRD pattern for the AgNP. The PXRD pattern was collected between the angles of 2θ from 3° to 90°.

3.1.5. BET-specific surface area measurements

Brunauer–Emmet–Teller (BET) surface areas were determined using ASAP 2010 (Accelerated Surface Area and Porosimetry System; Micromeritics Instrument Corporation, GA, USA). Prior to surface area analysis, samples were heated to 100°C and degassed overnight.

3.2. Specimen collection and acclimatization

Adult *P. perlatus* samples were collected from an uncontaminated area of the Eerste River (Stellenbosch, South Africa) during spring 2014. Individuals were collected using handmade traps comprising of a fishing rod fitted with mesh net containing bait. Capturing the individuals during spring ensured that they had not been exposed to heat wave conditions prior to their collection, thereby assuring that their recent thermal history did not include exposures to high temperatures. After capture, *P. perlatus* (with a mean length of 50 ± 5 mm and weight of 75 ± 10 g) were transported to the laboratory where they were kept unfed in aquaria (21 ± 2°C), and at a natural photoperiod for three days to acclimatise before exposure experiments began.

3.3. Experimental setup

In this study, laboratory experiments were conducted to investigate the physiochemical properties and temperature-dependent solubility of AgNPs that potentially influence their toxicity in aqueous environments and to aquatic organisms.

Details of the experimental setup are summarised in Table 1. The experimental setup comprised three experiments: (A) AgNP characterization in the dry state and in solution, (B) acute toxicity study, and (C) *in vivo* study. Details of each experiment are given in the sections below. Upon arrival in the laboratory, the crabs were kept in plastic aquaria composed of 2-L tanks and allowed to acclimatize for three days. There were six individuals per tank (10 systems with 6 individuals per tank = total of 60 crabs). Crabs were kept unfed during the acclimatization and exposure periods. Crabs were exposed for seven days. Ethical clearance was obtained and ethical animal care guidelines were followed.

	Nanoparticle characterization			In vivo study - Experiment A			Acute toxicity test - Experiment B		
	A.1	**A.2**	**A.3**	**B.1**	**B.2**	**B.3**	**C.1: AgNP - dependant**	**C.2: Temperature - dependant**	**C.3**
AgNP conc. (µg/gL)	1	1	1	0, 10, 100	0, 10, 100	0, 10, 100	0, 10, 100, 1 000, 10 000	782.77	782.77
Temp. (°C)	21	18	28	18, 18, 18	21, 21, 21	28, 28, 28	21 °C (RT)	16, 18, 22, 26, 28	25.37
Number of individuals (n)	6	6	6	6, 6, 6	6, 6, 6	6, 6, 6	6, 6, 6, 6, 6	6, 6, 6, 6, 6	6

Table 1. Experimental conditions for evaluating the effect of AgNPs (RT = room temperature)

3.3.1. *Silver nanoparticle characterization (Experiment A)*

Experiment A investigated the behaviour of AgNP (1 µg/mL) in three laboratory microcosms: (1) a control regime kept at room temperature (i.e. 21°C; A,1), (2) a low-temperature regime (i.e. 18°C; A.2) and (3) a high-temperature regime (i.e. 28°C; A.3).

3.3.2. Acute toxicity study (Experiment B)

The acute exposure study consisted of three experimental stages. Stage 1 involved a concentration-dependent regime comprised of crab specimens exposed to five different AgNP concentrations, including a control regime (i.e. 0, 1, 10, 100, 1000 and 10,000 μg/mL AgNPs), in order to cover a wide range of contamination levels that may be reported for polluted environments. A total of six crabs per treatment were exposed for seven days to room temperature. At the start of each day, the numbers of live and dead crabs were determined via visual inspection and the benchmark dose (BMD) AgNP concentration was derived through LogProbit analysis (U.S. EPA BMDS Program, version 2.5).

During stage 2, a total of six crabs per treatment were exposed for seven days to the final BMD AgNP (7 days) concentration (obtained in stage 1) at five pre-determined temperature regimes (i.e. 16°C, 18°C, 22°C, 26°C and 28°C). The experimental temperatures were chosen, taking into account the predicted increases in mean atmospheric and aquatic water temperatures (Bates et al.), since climate change projections indicate an increase in the frequency, intensity and duration of thermal extremes [29]. At the start of each day, the numbers of live and dead crabs were determined via visual inspection and the Critical thermal maximum (CTMax) were derived through LogProbit analysis (U.S. EPA BMDS Program, version 2.5) and was used to estimate the temperature to be used in stage 3. The CTMax is defined as "arithmetic mean of the collective thermal points at which the endpoint is reached" [30], or that temperature for a given species above which most individuals respond with unorganized locomotion, subjecting the animal to likely death [31].

Stage 3 involved the assessment of the role of oxidative stress in AgNP-induced toxicity, a total of six crabs per treatment were exposed to the corresponding BMD (782.77 μg/mL) and CTMax (25.37°C) values obtained during the preceding experimental stages. The experiment was conducted in 2-L plastic tanks (Group B1: six crabs at 0, 1, 10, 100, 1000 and 10,000 μg/mL at 21°C (room temperature) and Group B.2: six crabs at 782.77 μg/mL at 25.37°C) with a 12-h alternating light/dark cycle following modified methods described by Cheng [32]. Tissue samples for biochemical analysis were collected from each crab specimen at the end of Stage 3.

For all experimental stages, crabs were exposed for seven days and were unfed during the acclimatization and exposure periods. Every 24 h during stages 1 and 2, live crabs were counted and the dead crabs were removed. Death was assumed when no movement occurred when mechanically stimulated. No food was provided during the exposure period.

3.3.3. In vivo study (Experiment C)

The experimental temperatures for experiment A were 18°C, 21°C (room temperature) and 28°C. These temperatures were chosen to reflect the predicted increases in mean atmospheric and aquatic water temperatures [33]. Crabs were divided into three temperature-dependent regimes (i.e. 18°C, 21°C (control temperature) and 28°C) each containing three AgNP-dependent regimes (i.e. Group C.1 at 18°C: 0 μg/mL, 10 μg/mL and 100 μg/mL AgNPs; Group C.2 at 21°C: 0 μg/mL, 10 μg/mL and 100 μg/mL AgNPs and Group C.3 at 28°C: 0 μg/mL, 10 μg/mL and 100 μg/mL AgNPs). There were six individuals per regime (nine regimes with six crabs each = 54 crabs in total).

3.4. Tissue preparation

At the end of the exposure period, crabs were cryoanaethesized and tissues (gills and hepatopancreas) were excised from each crab sample. Tissues (gills and hepatopancreas) were homogenized (1:10 w/v) using an Omni-Ruptor 400 (Omni International Inc., GA, USA) homogenizer in a 1:20 protease inhibitor cocktail (Sigma Aldrich, MO, USA) prepared with a phosphate buffer (PBS). Homogenates were centrifuged (Universal 32R, Hettich Zentrigugen, Germany) at 4°C for 2 min at 13,000 rpm. Supernatants were removed and used for determination of enzyme activity.

3.5. Biochemical analysis

Biochemical analyses were done using commercially available kits purchased from a local supplier and were performed in triplicate following the manufacturer's protocols.

The enzymatic assay of SOD activity (Sigma-Aldrich, MO, USA), using nitroblue tetrazolium (NBT) and xanthine oxidase (XO), was carried out by reading the absorbance at 450 nm.

The enzymatic assay of CAT (Arbor Assays, MI, USA) was measured by following the increase in catalase in the sample with decreasing H_2O_2 concentration and by measuring the absorbance of 200 mM H_2O_2 at 520 nm. The reaction mixture consists of colorimetric detection reagent, horseradish peroxidase concentrate and H_2O_2.

The GST assay (Sigma-Aldrich, MO, USA) was measured at 340 nm following the conjugation of GSH with CDNB (1-chloro-2,4-dinitrobenzene). The reaction mixture consisted of Dulbecco's phosphate-buffered saline (DPBS), 200 mM L-glutathione reduced and 100 mM CDNB. The GST activity was measured by measuring the change in absorbance every minute for six minutes.

For normalization purposes, the results were divided by the total amount of protein (expressed in mg/g wet mass of tissue), calculated through the Bradford method (Bradford, 1976), to obtain enzymes activity in SOD units per mg protein (for SOD), CAT units per mg protein (for CAT) and GST-specific activity per mg protein (for GST).

3.6. Data integration

All data values are given as the mean ± SEM (standard error of means). Univariate one-way ANOVA was used to compare means between treatments followed by Dunnett's test to discriminate differences from the control group using XLSTat (Microsoft Excel and XLStat2015®). A minimum significance level of $p < 0.05$ was accepted. Benchmark dose (BMD) and the critical thermal maximum (CTMax) were calculated with LogProbit analysis (U.S. EPA BMDS Program, version 2.5).

4. Results and discussion

Although beneficial, advances in nanotechnology are also associated with expectations of growing potential toxicity and ecotoxicity largely due to their unusual properties. These properties that render NPs suitable in numerous applications are also ultimately the same properties responsible for unpredictable effects in the environment and produce adverse cellular effects and damage to living organisms. For example, small particle size and high specific surface area allows for higher reactivity and use in several applications, but also allows for their passage across biological barriers thereby entering cells [2,34]. It is therefore vital that NPs are correctly and accurately characterized in environmental media in order to ensure the reliability and reproducibility of toxicity tests.

Silver NPs are currently very widely used in industry largely due to their antibacterial properties, with applications in several consumer products. Once released into the environment, the state and behaviour of NP in the environment is dependent on environmental conditions (including temperature, pH, and ionic strength) in which the NP occurs. As such, characterization of both the physical and chemical properties of NM and that of the environment is necessary in order to predict the NM's behaviour and potential effects on the environment.

Commercial AgNPs were analysed in the dry state (TEM, SEM, BET, PXRD) to characterize surface composition and coatings, surface area, agglomeration state and size of the nanoparticles, in solution (TEM) to characterize particle size and aggregation potential in solution. The primary and aggregate size of the AgNPs in the dry state was characterized using TEM analysis (Figure 4a). A particle size distribution was generated from the TEM images by measuring the diameter of more than 500 nanoparticles (Figure 4b). The TEM image in Figure 4a verified the spherical nature of the AgNPs, while the size distribution histogram (Figure 4b) showed a majority of smaller particles measuring 10 nm and a small quantity of larger particles 30–50 nm. The image also shows that the smaller particles are primarily isolated but form agglomerates with diameters of 20 nm.

In the aquatic environment, AgNPs interact with natural water components, which can lead to chemical or biological alterations (such as size distribution, aggregation or disaggregation. This, in turn, will influence the potential transport of NPs in the water column and, consequently, their fate and toxicity. TEM images for each experimental regime (control, low and high temperature) are shown in Figure 5. It is known that larger aggregates imply reduced bioavailability and toxicity [35]. With this said, the enlarged aggregates evidenced in the TEM images at 18°C suggest that NP toxicity could be somewhat reduced. However, the TEM image at 28°C suggests a greater potential for toxicity since aggregation of the AgNPs are minimal. The propensity of NPs to aggregate in the aquatic environment can lead to sedimentation, thereby making NPs more available to interact with sediment-dwelling and benthic organisms than with pelagic (water dwelling) species.

The bioavailability of AgNPs in water and sediment in the different treatment regimes was assessed at the end of the exposure period (i.e. seven days). The available concentration of Ag was found to decrease with increasing temperature (4.7% at 18°C versus 1.1% at 28°C). The loss of Ag suggests aggregation, sedimentation and dissolution of the particles in the exposure media that resulted in lower AgNP concentrations in the water column and reduced bioavailability of AgNP to *P. perlatus*. The bioavailability of Ag is a critical element influencing its toxicity to aquatic organisms and is dependent on several factors such as aggregation, sedimentation and internalization within the test organism. With this said, it can be assumed that, from the TEM image at 18°C, AgNPs formed large aggregates, whereas at 28°C, aggregation was minimal. The lack of obvious aggregation on the water surface in the high-temperature regime likely reflects the internalization of AgNP at elevated temperatures and possible sedimentation.

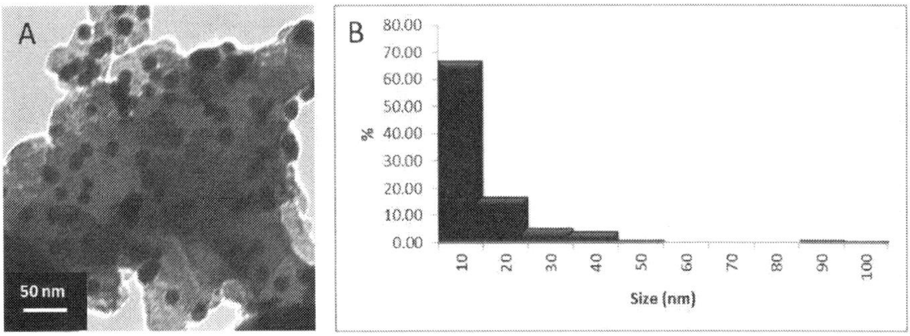

Figure 4. TEM image of dry AgNP (left) with the corresponding particle size distribution (right).

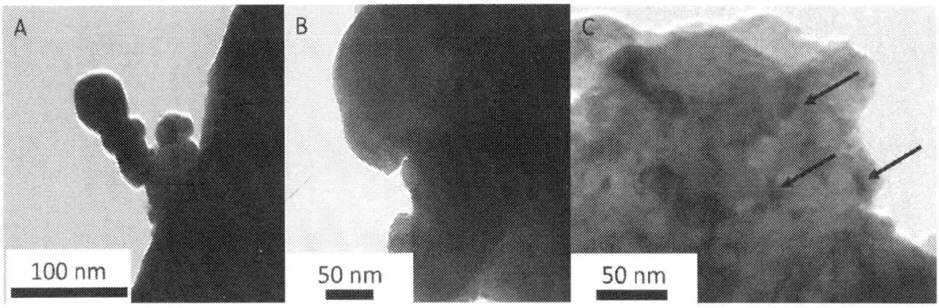

Figure 5. TEM AgNPs suspensions at (a) room temperature, (b) low temperature and (c) high temperature.

Silver NPs are capable of causing acute toxicity in *P. perlatus*; however, the toxicity differed significantly according to AgNP concentration and exposure temperature. The results of the seven-day acute toxicity tests performed with AgNPs and temperature, expressed as BMD values, are summarized in Table 2. In the acute toxicity test, survival in the control group was well above 90%, and thus met biological validity criterion. In Experiment B.1 (AgNP-depend-

http://dx.doi.org/10.5772/61715

ent regime), no mortalities were observed between 0 and 100 μg/mL AgNP. Mortalities were only observed in the 1000 μg/mL and 10,000 μg/mL treatment regimes, where mortality was observed from day 2 and increased until 75% and 100% mortality was confirmed at the end of the exposure period for the 1000 μg/mL and 10,000 μg/mL groups, respectively. In Experiment B.2 (temperature-dependent regime), 50% mortality was observed after 2 days at 28°C (Figure 4). No mortalities were observed in the 18°C and 22°C temperature groups. No significant differences in mortality between the two experimental stages were found. At the end of the exposure period, 75% had died in the temperature-dependent experiments. The mortality data during the experimental periods indicate that the AgNPs and temperature combinations were toxic to the survival of *P. perlatus*.

		SOD		CAT		GST	
Temperature (°C)	**AgNP (μg/mL)**	**G**	**HP**	**G**	**HP**	**G**	**HP**
Acute toxicity study (Experiment B)							
25.37	0	570.6 ± 18.0	203.5± 14.1	1842.8 ± 41.4	278.3 ± 41.4	89.6 ± 10.3	58.3 ± 10.3
25.37	787.77	1005.5 ± 136.3	233.6± 15.1	3490.9 ± 2468.5	388.7 ± 0.4	106.3 ± 19.9	42.8 ± 27.0
In vivo study (Experiment C)							
18	0	6698.4 ± 1221.7	3797.3 ± 249.5	896.5 ± 136.7	3269.8 ± 118.4	10426.2 ± 2858.0	11679.1 ± 1288.0
	10	4119.1 ± 535.0	3286.9 ± 693.9	977.7 ± 56.9	2989.2 ± 49.7	9066.5 ± 1975.7	21783.6 ± 4279.9
	100	2924.6 ± 591.7	4788.3 ± 740.2	684.3 ± 40.2	2327.9 ± 351.3	3826.1 ± 1368.2	12187.4 ± 1807.2
21	0	3135.4 ± 655.8	4357.4 ± 646.1	26061.6 ± 5963.0	3825.6 ± 1064.7	1683.9 ± 278.0	9802.4 ± 3540.7
	10	2728.4 ± 261.3	6584.0 ± 1284.9	14350.5 ± 3145.1	4016.8 ± 2150.3	1569.6 ± 267.8	17188.5 ± 5827.2
	100	3071.7 ± 250.4	6304.3 ± 756.0	10732.3 ± 835.3	3496.9 ± 2022.3	1669.7 ± 386.1	13342.8 ± 4038.7
28	0	3445.0 ± 836.3	3968.8 ± 673.2	20028.0 ± 6017.6	5184.7 ± 1301.1	11890.4 ± 2221.6	15018.7 ± 2708.8
	10	6458.3 ± 519.0	5297.3 ± 685.6	13526.1 ± 3048.6	13884.9 ± 2291.4	13865.6 ± 1127.6	40822.1 ± 6828.5
	100	9570.0 ± 1989.6	8339.1 ± 796.1	17765.2 ± 5873.4	22142.6 ± 2078.8	11197.1 ± 1483.0	48342.8 ± 5463.5

Table 2. Activities of antioxidant enzymes SOD (SOD units per mg protein), CAT (CAT units per mg protein)and GST (GST specific activity per mg protein)in tissues (G = gills; HP = hepatopancreas) of *P. perlatus* exposed to AgNP (10 μg/mL and 10 μg/mL) Data are presented as mean ± SEM (n = 6). Statistical significance (indicated by *) was denoted by $p < 0.05$ versus the respective control crabs.

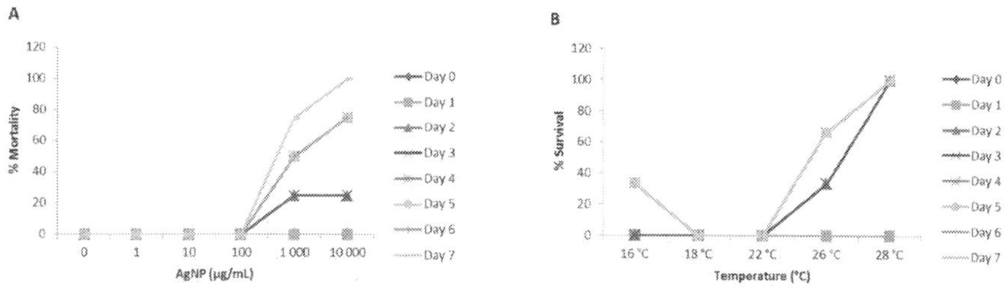

Figure 6. Mortality of *P. perlatus* exposed to various concentrations of AgNP (A) and various temperatures (B).

Based on these results, AgNPs appear to exert increased toxicity with increasing AgNP concentration and temperature. Previous studies have reported the toxic impacts of different NPs on crustaceans. In a recent study on *Chlamydomonas reinhardtii*, Navarro et al. reported that AgNPs (10–200 nm) induced 2 h EC_{50} of 3300 ± 572 nM [35]. Others have reported 48 h EC_{50} of 2.5 µg/mL and 4.9 µg/mL AgNPs in *Oncorhynchus mykiss* [36]. As such, these results suggest that AgNPs can generate different degrees of toxicity under different exposure conditions (such as NP size, coating and concentration, temperature and salinity) [37]. As such, these results suggest that AgNPs can generate different degrees of toxicity under different exposure conditions (such as NP size, concentration and temperature).

Oxidative stress is an important component of the stress responses in aquatic organisms, which are often exposed to a wide variety of environmental stressors (such as temperature variations and anthropogenic contamination). Biomarkers of oxidative stress are among the most commonly used biomarkers of cellular stress. These include superoxide dismutase (SOD), catalase (CAT) and glutathione-S-transferase (GST), which are used mainly as indicators of cellular stress resultant from both environmental contamination and environmental variables.

The crustacean antioxidant system is based on the collective actions of SOD, CAT and GST to counteract the overproduction of ROS. An important objective of this research was to assess the differential responses among different tissues. Because different tissues are located in different locations and perform different functions, each tissue may be more or less exposed to contaminants and will therefore have different cellular responses. To date, the gills and hepatopancreas have been the most used tissues for the determination of oxidative stress in crustaceans. The hepatopancreas is the major site for toxicant uptake and oxyradical-generating biotransformation enzymes [38]. The gills have a large exchange area and are in direct contact with the external environment [38]. For this reason, these tissues were chosen to assess the antioxidant defence mechanisms following AgNP and temperature exposures. In this study, in response to AgNP concentrations (i.e. 10 µg/mL and 100 µg/mL), antioxidant enzymes (SOD, CAT and GST) activity were significantly higher in the hepatopancreas when compared to the gills, suggesting that the hepatopancreas might be a more sensitive organ to the effects of AgNPs. Similarly, Zhu et al. [39] reported significant stimulation of SOD activity in liver (when compared to the gill and brain) following exposure of juvenile carp to C60. This indicates that AgNPs have tissue-specific effects on redox metabolism in *P. perlatus*.

Enzyme activity in the higher AgNP treatment (i.e. 100 μg/mL) was generally lower than the lower AgNP treatment (i.e. 10 μg/mL). SOD is the first defence against oxidative toxicity at a cellular level and is responsible for catalysing the dismutation of the superoxide radical O_2^- to O_2 and H_2O_2. The inductions of SOD and CAT in the higher-temperature and AgNP regimes suggest the production of superoxide anions by AgNP. The depletion of the antioxidant enzyme capacity (SOD, CAT and GST) suggests that the antioxidant defence system is overwhelmed by ROS [40] and further suggests that the antioxidant defence systems of these tissues were being stressed.

Experiment	Exposure (d)	BMD (μg/L) AgNP	95% confidence limit	Slope	Correlation coefficient
Experiment B.1					
	24	-	-	-	-
	48	3689.16	18 827.00	$p < 0.05$	0.29
	72	3689.16	18 827.00	$p < 0.05$	0.29
	96	4209.83	1 947.73	$p < 0.05$	0.55
	120	4209.83	1 947.73	$p < 0.05$	0.55
	144	4209.83	1 947.73	$p < 0.05$	0.55
	168	4083.36	782.77	$p < 0.05$	2.75
Experiment	**Exposure (h)**	**CTMax (°C)**	**95% confidence limit**	**Slope**	**Correlation coefficient**
Experiment B.2					
	24	-	-	-	-
	48	26.14	$p < 0.05$	18.00	0.83
	72	26.14	$p < 0.05$	18.00	0.83
	96	25.37	$p < 0.05$	18.00	0.75
	120	25.37	$p < 0.05$	18.00	0.75
	144	25.37	$p < 0.05$	18.00	0.75
	168	25.37	$p < 0.05$	18.00	0.75

Table 3. BMD values of the performed acute toxicity tests with AgNPs and temperature on *P. perlatus*

Elevated temperature has been shown to result in oxidative stress and affect enzyme activity. As a consequence, the induction of antioxidant defences is an important component of the stress response against oxidative stress [41]. The results reported here generally support the fact that oxidative stress biomarkers are highly sensitive to temperature, likely due to temperature-induced ROS production. It is hypothesized that exposure to AgNPs and elevated temperature will generate ROS and elicit oxidative stress in *P. perlatus*. Oxidative stress responses are generally lowest at conditions of lowest stress (i.e. at 21°C—similar to room temperature) and highest at conditions of highest stress (i.e. 100 μg/mL AgNPs at 28°C and 18°C). In our study, antioxidant enzymes (SOD, CAT and GST) activities were generally lower at lower AgNP concentrations and at lower temperatures (e.g. at 0 μg/mL and 100 μg/mL, and at 18°C and 21°C), and was highest at higher AgNP concentrations and at higher temperature (e.g. at 100 μg/mL and at 28°C). Since the dissolution of oxygen is generally higher at lower temperatures (and vice versa), it is expected that oxidative stress responses are inversely correlated to temperature [42]. However, the present study indicates that oxidative stress responses were generally lowest at conditions of lowest stress (i.e. at 21°C and 18°C). Other studies have also provided similar evidence of temperature effects on oxidative stress parameters. For example, Vinagre et al. [43] reported the effect in tissues of *Gobius paganellus* in an experiment at increasing temperatures. In an earlier study, Vinagre et al. [42] also reported variation in oxidative stress responses (i.e. increased catalase activity) due to temperature (28°C) in *Dicentrarchus labrax*. A similar conclusion was reported by Rodrigues et al. [44] in in the muscle and digestive gland of *Callinectes maenas*, while Paital et al. [45] reported the seasonal effects on oxidative stress biomarkers in the gills and hepatopancreas of *Scylla serrata*. The capacity for aggregation, sedimentation and solubility of AgNPs in aqueous environments can limit their transport within the water column, and thereby reduce their bioavailability to aquatic organisms [46]. As seen, smaller aggregates were formed at the higher temperature regime, thereby enhancing the potential for toxicity.

5. Conclusions and significance

Nanoparticle toxicity is a growing concern in freshwater habitats. However, understanding NP effects on aquatic organisms is largely impeded by the lack of the studies addressing these effects combined with other environmentally relevant stressors. The present study was designed to investigate the behaviour of AgNPs in aqueous suspension under different environmental parameters with particular focus on environmental conditions such as temperature, and the concomitant effects on AgNP uptake, toxicity and antioxidant defence mechanisms in a freshwater crab species *P. perlatus*, common in the waters of southwestern region of the Western Cape [47]. Nanoproducts are increasingly being used in various products and, consequentially, the potential adverse effects associated with exposure to NMs are of concern. The risks associated with NMs (i.e. its fate and behaviour in the environment) are largely unknown and difficult to predict. As the ultimate sink for conventional contaminants, the aquatic ecosystem is therefore predisposed to the potential effects of NPs. Although our knowledge on the toxicity of various NMs in the aquatic environment has increased over the past few years, there is still a lack of knowledge regarding exposure concentrations, bioaccumulation in tissues, as well as environmental factors which could potentially affect its toxicity or bioaccumulation. This book chapter emerges in this context, centring on the effects of the most commonly used and commercially available AgNP using *P. perlatus* as a sentinel species.

Coexposure to AgNPs and elevated temperature resulted in a significant increase in ROS production rates and increase in antioxidant enzyme activity. Elevated temperatures often increase the negative impacts of pollutants in aquatic organisms, especially as the temperatures approach the upper tolerance limits. As such, it is imperative to assess the effects of pollutants (such as NPs) on sentinel species in the context of the environmentally relevant thermal variability.

6. Future perspectives

The work conducted here has focussed largely on the collective effects of AgNPs and temperature on the oxidative stress defences of *P. perlatus*. Whether the reported results could be applicable in a more environmentally realistic setting has to be investigated; however, these findings do offer several directions for future research:

- South Africa's National Nanotechnology Strategy (DST, 2007) envisages the exploitation of nanotechnology in South Africa. Future studies should consider the transformation of AgNPs (both coated and uncoated) as it passes through wastewater treatment plants. Since the levels of AgNPs are expected to increase in the environment, one should question whether our wastewater treatment plants are capable of managing the elevated levels in terms of their treatment capacity and efficiency. As such, the evaluation of the removal of selected NPs in wastewater by different water treatment processes should be undertaken in order to estimate the concentrations of NPs in reclaimed wastewater for potable reuse.

- As evidenced, AgNPs are not, at present, in isolation. It is therefore important that future studies consider the effects of multiple toxins such as emerging pollutants (pharmaceuticals, etc.). Other abiotic factors are also worthy of consideration in future coexposure studies.

- Previous studies have reported notable differences between the responses of male and female individuals, suggesting that there may be some gender-specific effects to NP exposure. Different responses between genders should thus be considered.

- Future investigations should assess the combined oxidative stress responses of AgNPs and lower temperature limits, i.e. critical thermal minima (CTMin).

Acknowledgements

This research was supported by the CSIR. The authors wish to thank Christoff Truter (University of Stellenbosch), Dewald Schoeman (University of the Western Cape), Charlton van der Horst (CSIR) and Sebastian Brown (CSIR, Chemical and Analytical Services, Stellenbosch) for assisting in various aspects of the research.

Author details

Chavon Walters[1], Edmund Pool[2] and Vernon Somerset[1]

*Address all correspondence to: cwalters@csir.co.za

1 Council for Scientific and Industrial Research (CSIR), Stellenbosch, South Africa

2 Department of Medical Biosciences, University of the Western Cape (UWC), Bellville, South Africa

References

[1] Dowling A. Development of nanotechnologies. Materials Today. 2004;7:30–35.

[2] Scown TM, Santos E, Johnston BD, Gaiser B, Baalousha M, Mitov S, Lead JR, Stone V, Fernandes T, Jepson M, van Aerle R, Tyler CR. Effects of aqueous exposure to silver nanoparticles of different sizes in rainbow trout. Toxicological Sciences. 2010;115(2): 521–534.

[3] Piccinno F, Gottschalk F, Seeger S, Nowack B. Industrial production quantities and uses of ten engineered nanomaterials in Europe and the world. Journal of Nanoparticle Research. 2012;14:1109–1119.

[4] Blaser SA, Scheringer M, MacLeod M, Hungerbühler K. Estimation of cumulative aquatic exposure and risk due to silver: Contribution of nano-functionalized plastics and textiles. Science of the Total Environment. 2008;390:396–409.

[5] Luoma, S.N., Silver Nanotechnologies and the Environment: Old Problems or New Challenges. Woodrow Wilson International Centre for Scholars and PEW Charitable Trusts, Washington, DC. : Project on Emerging Nanotechnologies. Publication 15. ; 2008.

[6] Ahamed M, Posgai R, Gorey TJ, Nielsen M, Hussain SM, Rowe JJ. Silver nanoparticles induced heat shock protein 70, oxidative stress and apoptosis in *Drosophila melanogaster*. Toxicology and Applied Pharmacology. 2010;242:263–269.

[7] Levard C, Hotze EM, Lowry GV, Brown GE. Environmental transformations of silver nanoparticles: impact on stability and toxicity. Environmental Science and Technology. 2012;46:6900–6914.

[8] Piao MJ, Kang KA, Lee IK, Kim HS, Kim S, Choi JY, Choi J, Hyun JW. Silver nanoparticles induce oxidative cell damage in human liver cells through inhibition of reduced glutathione and induction of mitochondria-involved apoptosis. Toxicology Letters. 2011;201:92–100.

[9] Fabrega J, Fawcett SR, Renshaw JC, Lead JR. Silver nanoparticle impact on bacterial growth: effect of pH, concentration, and organic matter. Environmental Science and Technology. 2009;43:7285–7290.

[10] Miao AJ, Schwehr KA, Xu C, Zhang SJ, Luo ZP. The algal toxicity of silver engineered nanoparticles and detoxification by exopolymeric substances. Environmental Pollution. 2009;157:3034–3041.

[11] Halliwell B, Gutteridge JM. Free Radicals in Biology and Medicine. 4th ed. New York: Oxford University Press; 1999. 704 p.

[12] Chitra K.C, Sajitha R. Effect of bispenol-a on the antioxidant defense system and its impact on the activity of succinate dehydrogenase in the gill of freshwater fish, *Oreochromis mossambicus*. Journal of Cell & Tissue Research. 2014;14(2):4219–4226.

[13] Falfushynska H, Gnatyshyna L, Yurchak I, Sokolova I, Stoliar O. The effects of zinc nanooxide on cellular stress responses of the freshwater mussels *Unio tumidus* are modulated by elevated temperature and organic pollutants. Aquatic Toxicology. 2015;162:82–93.

[14] Bucheli TD, Fent K. Induction of cytochrome P450 as a biomarker for environmental contamination in aquatic ecosystems. Critical Reviews of Environmental Science and Technology. 1995;25:201–268.

[15] Almroth BC.. Oxidative Damage in Fish Used as Biomarkers in Field and Laboratory Studies [thesis]. University of Gothernburg:2010. 74.

[16] Kil IS, Park J-W. Regulation of mitochondrial NADP+-dependent isocitrate dehydrogenase activity by glutathionylation. Journal of Biological Chemistry. 2005;280:10846–10854.

[17] USEPA (United States Environmental Protection Agency). Nanomaterial Case Study: Nanoscale Silver in Disinfectant Spray. EPA/600/R-10/081F; 2012. 423.

[18] Asghari S, Johari SA, Lee JH, Kim YS, Jeon YB, Choi HJ, Moon MC, Yu IJ. Toxicity of various silver nanoparticles compared to silver ions on *Daphnia magna*. Journal of Nanobiotechnology. 2012;10:14–34.

[19] Lorenz MR, Holzapfel V, Musyanovych A, Nothelfer K, Walther P, Frank H. Uptake of functionalized, fluorescent-labeled polymeric particles in different cell lines and stem cells. Biomaterials. 2006;27:2820–2828.

[20] Jayaseelan C, Rahuman AA, Ramkumar R, Perumal P, Rajakumar G, Kirthi AV, Santhoshkumar T, Marimuthu S. Effect of sub-acute exposure to nickel nanoparticles on oxidative stress and histopathological changes in Mozambique tilapia, *Oreochromis mossambicus*. Ecotoxicology and Environmental Safety. 2014;107:220–228.

[21] Moore MN. Do nanoparticles present ecotoxicological risks for the health of the aquatic environment?. Environmental International. 2006;32:967–976.

[22] Zhao CM, Wang WX. Biokinetic uptake and efflux of silver nanoparticles in *Daphnia magna*. Envionmental Science and Technology. 2010;44:7699–7704.

[23] Fabrega J, Luoma, SN, Tyler CR, Galloway TS, Lead JR. Silver nanoparticles: behaviour and effects in the aquatic environment. Environmental International. 2011;37:517–513.

[24] Cumberlidge N. The freshwater crabs of West Africa: family Potamanautidae. IRD; 1999. 382.

[25] Qin Q, Qin S, Wang L, Lei W. Immune responses and ultrastructural changes of hemocytes in freshwater crab *Sinopotamon henanense* exposed to elevated cadmium. Aquatic Toxicology. 2012;106–107:140–146.

[26] Issartel J, Boulo V, Wallon S, Geffard O, Charmantier G. Cellular and molecular osmoregulatory responses to cadmium exposure in *Gammarus fossarum* (Crustacea, Amphipoda). Chemosphere. 2010;81:701–710.

[27] De Freitas Rebelo M, Rodriguez EM, Santos EA, Ansaldo M. Histopathological changes in gills of the estuarine crab *Chasmagnathus granulata* (Crustacea-Decapoda) following acute exposure to ammonia. Comparative Biocehmistry and Physiology, Part C. 2000;125:157–164.

[28] Lee RF. Bioavailability, biotransformation and fate of organic contaminants in estuarine animals. In: Eds. Newman, M.C., Roberts, M.H. Jr and Hale,, editors. Coastal and Estuarine Risk Assessment. CRC Press, R.C. Lewis Publishers; 2001. p. 368.

[29] IPCC (Intergovernmental Panel for Climate Change). Third assessment report of the working group I. In: Houghton, J.T., et al., editors. The Science of Climate Change. 3rd ed. Cambridge, United Kingdom and New York, NY, USA: Cambridge University Press; 2001. p. 881.

[30] Mora C, Ospina A. Tolerance to high temperatures and potential impact of sea warming on reef fishes of Gorgona Island (tropical eastern Pacific). Marine Biology. 2001;139:765–769.

[31] McDiarmid RW, Altig R. Tadpoles: The Biology of Anuran Larvae. University of Chicago Press ; 1999. 444.

[32] Cheng PW, Liu SH, Hsu CJ, Lin-Shiau SY. Correlation of increased activities of Na+, K+-ATPase and Ca2+-ATPase with the reversal of cisplatin ototoxicity induced by D-methionine in guinea pigs. Hearing Research. 2005;205:105–109.

[33] Bates BC, Kundzewicz ZW, Wu S, Palutikof JP, editors. Climate Change and Water. Technical Paper of the Intergovernmental Panel on Climate Change. Geneva: IPCC Secretariat; 2008. 210.

[34] Handy RD, Henry TB, Scown TM, Johnston BD, Tyler CR. Manufactured nanoparticles: their uptake and effects on fish—a mechanistic analysis. Ecotoxicology. 2008;17:396–409.

[35] Navarro E, Piccapietra F, Wagner B, Marconi F, Kaegi R, Odzak N. Toxicity of silver nanoparticles to *Chlamydomonas reinhardtii*. Environmental Science and Technology. 2008;42:8959–8964.

[36] Farkas J, Christian P, Urrea JA, Roos N, Hassello°v M, Tollefsen KE, Thomas KV. Effects of silver and gold nanoparticles on rainbow trout (*Oncorhynchus mykiss*) hepatocytes. Aquatic Toxicology. 2009;96:44–52.

[37] Yang X, Gondikas AP, Marinakos SM, Auffan M, Liu J, Hsu-Kim H, Meyer JN. Mechanism of silver nanoparticle toxicity is dependent on dissolved silver and surface coating in *Caenorhabditis elegans*. Environmental Science and Technology. 2012;46:1119–1127.

[38] Lavarias Sl, Heras H, Pedrini N, Tournier H, Ansaldo M. Antioxidant response and oxidative stress levels in *Macrobrachium borellii* (Crustacea: Palaemonidae) exposed to the water-soluble fraction of petroleum. Comparative Biochemistry and Physiology, Part C. 2011;153:415–421.

[39] Zhu XS, Zhu L, Lang YP, Chen YS. Oxidative stress and growth inhibition in the freshwater fish *Carassius auratus* induced by chronic exposure to sublethal fullerene aggregates. Environmental Toxicology and Chemistry. 2008;27(9):1797–1985.

[40] Vander O R, Beyer J, Vermeulen N P E. Fish bioaccumulation and biomarkers in environmental risk assessment: A review. Environmental Toxicology and Pharmacology. 2003;13:57–149.

[41] Parihar MS, Javeri T, Hemnani T, Dubey AK, Prakash P. Responses of superoxide dismutase, glutathione peroxidase and reduced glutathione antioxidant defences in gills of the freshwater catfish (*Heteropneustes fossilis*) to short-term elevated temperature. Journal of Thermal Biology. 1997;22:151–156.

[42] Vinagre C, Madeira D, Narciso L, Cabral H, Diniz M. Effect of temperature on oxidative stress in fish: lipid peroxidation and catalase activity in the muscle of juvenile seabass, *Dicentrarchus labrax*. Ecological Indicators. 2012;23:274–279.

[43] Vinagre C, Madeira D, Mendonca V, Dias M, Roma J, Diniz M. Effect of increasing temperature in the differential activity of oxidative stress biomarkers in various tissues of the rock goby, *Gobius paganellus*. Marine Environmental Research. 2014;97:10–14.

[44] Rodrigues AP, Oliveira PC, Guilhermino L, Guimarães L. Effects of salinity stress on neurotransmission, energy metabolism, and anti-oxidant biomarkers of *Carcinus maenas* from two estuaries of the NW Iberian Peninsula. Marine Biology. 2012;159:2061–2074.

[45] Paital B, Chainy GBN. Seasonal variability of antioxidant biomarkers in mud crabs (*Scylla serrata*). Ecotoxicology and Environmental Safety. 2013;87:33–41.

[46] Baalousha M, Manciulea A, Cumberland S, Kendall K, Lead JR.. Aggregation and surface properties of iron oxide nanoparticles; influence of pH and natural organic matter. Environmental Toxicology and Chemistry. 2008;27:1875–1882.

[47] Snyman R G, Reinecke A J, Nel JAJ. Uptake and distribution of copper in the fresh-water crab, *Potamonautes perlatus* (Crustacea) in the Eerste River, South Africa. South African Zoology. 2002;81–89.

Chapter 3

Overview of the Standard Methods for Soil Ecotoxicology Testing

Paulo Roger Lopes Alves and Elke Jurandy Bran Nogueira Cardoso

Additional information is available at the end of the chapter

http://dx.doi.org/10.5772/62228

Abstract

This chapter briefly describes the importance of the services provided by soil invertebrates in terrestrial ecosystems and highlights the role of soil fauna in the risk assessments of potentially polluting substances for the terrestrial environment, considering the sensitivity of these organisms, when compared to other indicators of soil quality (e.g., chemical and physical). The main invertebrate groups used in laboratorial ecotoxicological assays are presented and, based on its physiological characteristics and habit requirements, the advantages and disadvantages of using certain taxonomic groups in laboratory assessments are also discussed. The most frequently used methods to perform this type of toxicity tests are summarized, highlighting the fundamental steps of the assays with the species *Eisenia fetida/Eisenia andrei, Folsomia candida, Enchytraeus albidus/Enchytraeus crypticus,* and *Hypoaspis aculeifer,* as well as the possible adjustments that are being carried out in tropical countries. Finally, the future prospects, related to the challenge of increasing the realism of laboratory ecotoxicological analyses, are discussed to show the main needs of this study at global and regional perspectives.

Keywords: Soil invertebrates, Bioassays, Ecotoxicity, Risk assessment, Standard procedures

1. Introduction

1.1. The role of soil invertebrates in soil-risk assessments

Soil represents one of the most complex and diverse ecosystems on earth. In addition to providing the main environmental support for the majority of plants, soil provides the habitat for a vast diversity of animals (vertebrate and invertebrate) and microorganism taxa. Soil is estimated to harbor one-fourth of all of the described biodiversity [1–2]. Although organisms may occur in almost all soil profiles, biological activity is highly concentrated in the most superficial layers, mainly within the top 30 cm of soil, where the highest concentrations of organic material are also found [3–4].

The transformations promoted by soil organisms, especially those that benefit human populations, are known as ecosystem services [5]. The concept of ecosystem services is recent, and a large number of services have already been identified as priorities for terrestrial ecosystems because they significantly influence the daily life of human populations and are fundamental for the maintenance of ecosystems and agriculture.

Soil invertebrates are directly or indirectly responsible for various biological and biochemical processes underlying terrestrial ecosystem services, namely, at the level of microbiota regulation, nutrient cycling, soil structuring, and water-quality regulation through filtration processes. These organisms act directly on the fragmentation and distribution of organic material deposited in the soil (animal and plant debris) and function as catalysts of organic matter (OM) decomposition processes and the biogeochemical cycles of carbon, nitrogen, phosphorus, and sulfur [5]. Moreover, these organisms regulate microbial activity (including the control of pathogens), microorganism distribution along the soil profile, and soil structuring because certain fauna species build tunnels, galleries, and other structures along soil profiles (horizontally and vertically) that are used to transport microorganisms and/or their latent forms (e.g., fungal spores) and provide preferential routes for organic matter incorporation and distribution as well as for root growth.

A classic example of a terrestrial ecosystem service provided by soil invertebrates is their influence on the mineralization of nutrients contained in the soil organic matter (SOM), a process that is fundamental for the maintenance of agriculture and forestry systems. This recycling of chemical elements is responsible for the supply of a large portion of the nutrients required by plants as well as increases in the productivity of these systems [6]. Therefore, it is essential to protect the ecosystem services provided by soil fauna.

In order to rapidly meet the high demand for food and products at lower production costs, methods that are more aggressive to the environment are being used, such as the intensive use of pesticides (to control pests and diseases), discharge of agro-industrial waste directly into the soil (without treatment), and other types of exploitation of the edaphic system [7]. These interventions may present a high risk for terrestrial ecosystems because they constitute entryways of several potentially toxic contaminants that may compromise invertebrate performance and their services, thus also affecting soil microorganisms and the functionality of soil ecosystems.

Until recently, the impact of anthropogenic contamination of terrestrial ecosystems has been primarily measured using indicators of soil chemical and physical properties because biological properties have generally been considered more difficult to predict or even measure [8]. The pH, cation-exchange capacity (CEC), organic matter content, and soil nutrient levels (fertility) are the most commonly used chemical parameters for evaluating soil health. However, these parameters are especially relevant when analyzing the soil's capacity to increase crop yield [9]. Similarly, the physical parameters texture, aggregate stability, soil density, and soil porosity, which are simple, fast and lowcostanalyzes [9], can be used as indicators of soil health. Nevertheless, they aremainly related to hydrological processes such as erosion, aeration, soil runoff, infiltration rate, and water-holding capacity (WHC) [10].

A large number of traditional studies that have applied environmental impact indicators for terrestrial ecosystems have only utilized the parameters organic matter (as chemical indicator) and soil-aggregate stability (as physical indicator), whereas a smaller number of studies have correlated biological indicators (bioindicators) to soil quality [8]. However, the impact of contaminants on soil may be more easily identified through their effects on biota than from the results of chemical and physical analyses [11–13] because of the higher sensitivity of biological processes and the capacity of organisms to detect and rapidly respond toa particular part of the contaminant concentrations in the soil (e.g., heavy metals), which is the fraction of contaminants available for uptake by living organisms known as the bioavailable fraction [14]. Therefore, assessing the impact of contaminants through their effects on biota may provide an early warning of risks to terrestrial ecosystems [3]. In addition, assessments using living soil organisms, called bioindicators, can be used to determine whether contaminants released into soil affect ecosystem services.

Standardized ecotoxicological laboratory tests using invertebrates are one of the first steps in risk assessments of soil contaminants, and they can be considered one of the main assessment tools because they are frequently sufficient for determining the ecological risk level of substances and the safe exposure limits for humans and soil biota [13, 15]. This type of test provides quantitative and/or qualitative information on the toxic effects of contaminants on soil invertebrate fauna, including information required by several global regulatory authorities prior to the sale of pesticides [16] or to allow the application of residues to agricultural soils [17].

In the European Union (EU), the sale of phytopharmaceuticals (including pesticides) is regulated by specific guidelines that require standardized ecotoxicological tests to assess the impact of these substances on nontarget soil organisms [16, 18]. Regarding the application of agro-industrial residues to soil, the EU requires that assessments be performed according to Directive 2008/98/EC [17], which includes an ecotoxicological risk criterion (H14 – "ecotoxic"). This criterion is used to identify wastes that constitute or may constitute immediate or delayed risks for one or more sectors of the environment (including soil invertebrates) and may be the determining factor used to decide the hazard level of more than 80% of wastes [19].

Two different approaches can be distinguished in laboratory ecotoxicological tests [20]. The first approach involves analyses that are predictive/prognostic and aim to determine the possible toxic effects of the substances tested on invertebrates in case the substances are released into the soil. This approach is mainly used to test new substances (e.g., new pesticides and pharmaceutical drugs) for which the safe exposure levels in terrestrial environments are unclear; thus, this approach can be used to regulate their use or prevent their introduction to the market. The second approach involves analyses that are diagnostic and aim to determine the actual ecological risk or current damage using samples of contaminated natural soils. In this case, the assessments are used to define the priorities for remediating contaminated areas and/or actions for reducing ecological risks.

These two approaches include various tests that can be classified according to exposure time (acute or chronic toxicity), observed effect (mortality, reduced growth or reproduction, bioaccumulation, or behavioral changes) or effective response (lethal or sublethal) [21]. In these tests, representative species of the soil fauna are exposed to increasing contaminant concentrations and the contaminant effects are measured in one (single species) or several species (multispecies) to test dose–response relations [22].

There are various methods for evaluating toxicity in invertebrates, including topical application, force-feeding, and immersion tests [23]. However, the main laboratory assays standardized by the norms of the Organization for Economic Cooperation and Development (OECD) and International Organization for Standardization (ISO) consist of exposing standard species to samples of contaminated soil. These protocols describe methods used to determine acute and chronic toxicity and the effects on the behavior of earthworms, collembolans, enchytraeids, mites, mollusks, and few other insects [20].

The objective of acute toxicity tests is mainly to assess whether a substance causes organism death. These tests are useful for short-term identifications of highly toxic contaminants; however, they do not consider different stages of the test organism life cycle (growth, reproduction, and birth of juveniles) or determine whether particular life cycle stages present increased sensitivity to toxic substances [24]. These tests are also used as preliminary evaluations ("range-finding tests") to determine the concentration ranges to be used in definitive acute toxicity tests and/or the sublethal concentrations for chronic toxicity assays [13].

Chronic toxicity tests are medium-term tests that measure the sublethal effects of potentially toxic substances on organisms, such as changes in reproduction and growth, and are more adequate for assessing effects at the population level [20, 25]. The primary standard methods for laboratory chronic toxicity tests have been established in ISO [26–27] and OECD [28–31] guidelines. The objective of these standardized tests is similar for different groups of invertebrates, although they do present differences, especially in test duration, as a result of the different reproductive characteristics of different species. In these tests, adults are exposed to a range of sublethal concentrations of the test substance, with the concentrations defined according to preliminary tests (range finding) or results from the literature.

Behavioral tests with soil invertebrates are also becoming common because they provide a preliminary evaluation of responses to soil pollution over a shorter period relative to that of toxicity tests [13]. In addition to providing ecologically relevant results because of the sensitivity of species in detecting polluting substances in soils, these tests can be performed more quickly (2 days on average) and at a reduced cost [13, 32]. Avoidance tests, for example, can be used as triage tools to assess the habitat suitability of soils because they are based on the ability of animals (e.g., earthworms) to avoid potentially toxic substances upon exposure to contaminated soils [33] because of the presence of chemoreceptors that are highly sensitive to chemicals in the environment [34].

Although behavioral assays offer alternative endpoints for assessing the impact of contaminants on soil invertebrates, it is recommended to use such evaluations along with acute and/or chronic toxicity tests [35] because in certain cases, the substances can cause 100%

mortality without the observance of an avoidance effect [32]. Such cases may be related to the test substance's narcotic properties, an absence of irritating effects, or physiological adaptations associated with the species' mode of life [36]. Integrated laboratory analyses should therefore be performed when a higher precision is required when assessing the interactions between contaminants, animal species, and soil properties because such analyses decrease the amount of uncertainty when determining ecological risks.

Although the standardized guidelines for ecotoxicological tests for terrestrial environments are relatively new, compared with the guidelines for aquatic environments, the number of tests based on these guidelines has increased considerably, and such tests have been used to investigate the ecological risk assessment of fungicides, herbicides, insecticides, heavy metals, nanomaterials, agro-industrial residues, and other substances in soil [13, 20].

2. Standard species for soil ecotoxicology

Ideally, the toxic effects of all chemicals introduced to terrestrial ecosystems, such as agro-industrial and urban wastes, should be tested on all species inhabiting the ecosystem before commercialization (xenobiotics) or direct application to the soil [13]. Because these measures cannot be achieved in the laboratory, edaphic invertebrate species that have known sensitivities to anthropogenic changes and provide the main representative ecosystem services of the fauna have been chosen as indicators of the ecotoxicological risk to terrestrial ecosystems. The review presented in Ref. [20] provides a list of the main invertebrates used in terrestrial ecotoxicological assays. This list includes earthworms (*Eisenia andrei* and *Eisenia fetida*), enchytraeids (*Enchytraeus albidus* and *Enchytraeus crypticus*), mollusks (*Helix aspersa*), mites (*Hypoaspis aculeifer, Platynothrus peltifer,* and *Oppia nitens*), isopods (*Porcellio scabere* and *Porcellionides pruinosis*), collembolans (*Folsomia candida* and *Folsomia fimetaria*), insects of the family Carabidae (*Pterostichus oblongpunctatus* and *Poecilus cupreus*), *Oxythyrea funesta,* and other organisms used in the methods standardized by international guidelines for ecological risk assessments of soil contaminants.

The methods described by the ISO and OECD international guidelines for ecotoxicity tests with terrestrial invertebrates are designed to standardize the tests so that similar results can be obtained in different laboratories regardless of the region. Such standardization facilitates comparisons and increases the reliability of the established toxicity levels. To develop standardized tests, standard species must be selected based on ecological relevance, ease of maintenance in the laboratory, and short-generation time [37–38]. In addition, the selected species should have well-known biological parameters so that a large number of healthy, homogeneous (same size and biomass), and age-synchronized individuals can be obtained. The number of species that meet all of these requirements is small.

In general, toxicity assays with standard species are performed individually; therefore, each species is tested separately to exclude the effects of interactions among species present in soil. However, the use of several species, even when tested separately, increases the ecological relevance of laboratory analyses because different organisms respond differently to pollutants, and the potential risk to the ecosystem varies [20]. The performance of a balanced battery of tests using organisms of different functional and taxonomic groups and different routes of exposure is therefore necessary to improve the reliability of ecological risk assessments determined via laboratory tests [39].

The main standard invertebrates used in batteries of soil ecotoxicological tests are earthworms [29, 40–42], collembolans [26, 31], mites [39], and enchytraeids [22, 28, 43]. Because of different morphological (e.g., epidermis structure) and physiological (e.g., water and oxygen uptake pathways) characteristics as well as feeding and behavioral habits (e.g., movement within the soil or at the soil surface and digging habits), these taxonomic groups encompass different pollutant uptake routes and encounter pollutants through the exposure to water and air present in the soil pores and ingestion of food and soil particles [44].

Earthworms are important components of the soil biota because they aid in the formation and maintenance of soil structure and fertility. Although they are not numerically dominant, their large size makes them one of the main contributors to invertebrate biomass in soil. These organisms are important indicators of soil life and quality because their populations are affected by common agricultural practices, especially by the use of pesticides and fertilizers and application of waste [45]. Because they are soft-bodied organisms, earthworms absorb water mainly through their skin; therefore, they can accumulate chemicals during water absorption [44]. Another important route of contaminant absorption is through ingestion because these organisms ingest large amounts of soil with adsorbed substances along with their food (soil organic material, which may also be contaminated).

Although several earthworm species have been used in terrestrial ecotoxicological tests [22], only *E. fetida* and *E. andrei* were included in the ISO and OECD guidelines. These species are preferred because they have worldwide distribution, are naturally tolerant to various organic substrates, are easily handled in single species or mixed cultures [45], and may be easily acquired commercially (adults, juveniles, or cocoons) or obtained from other soil ecotoxicology laboratories.

E. fetida and *E. andrei* were initially differentiated into two forms, var. tipica and var. unicolor (lightly stripped and uniformly pigmented, respectively), by Bouché (1972), who considered them as subspecies (*E. fetida fetida* and *E. fetida andrei*). These organisms were thought to belong to the same species for a long time because of their similar appearances, ecological demands, and frequent associations. However, recent investigations have determined that their crossing does not result in viable descendants, and biochemical methods were used to confirm that they constitute different species [46]. This differentiation is important for their use in ecotoxicological tests because the effects of contaminants on the two species may be different.

Neither of the two species is typical of most agricultural soils, and they only occur in soils rich in organic matter. Under ideal conditions, their life cycle until maturity is relatively short (varying between 45 and 51 days) compared with that of other species and extends from the recently deposited cocoon until the adult stage, when the worms are sexually mature (with the presence of clitellum) and ready to produce the next generation. The time for juveniles (recently hatched) to reach sexual maturity varies between 21 and 30 days. Both species are prolific, and between two and five juveniles are generated from each viable cocoon. Depending on the rearing temperature and substrate, their maximum life span ranges from four and a half to five years [45].

Enchytraeids (family Enchytraeidae) belong to the same class as earthworms (Oligochaeta) and can live in both water and soil. However, they are mostly found in soil, where they perform important ecosystem services, such as increasing the rate of organic matter decomposition, maintaining the soil structure (creation of biopores), and dispersing microbes on a local scale [47–48]. These services are especially important in acidic, sandy, and nutrient-poor soils, where enchytraeids are the dominant soil fauna group (up to 75% of the biomass) [48–49]. In these environments, the role of enchytraeids in organic matter decomposition may not be performed by other fauna groups [50].

The family Enchytraeidae has over 600 described species [48]. Known as white worms or potworms because of their pale color and small size (many are only a few millimeters long, although some may reach up to 5 cm), most enchytraeids are hermaphroditic (capable of self-fertilization), although certain species are parthenogenetic or reproduce through fragmentation [51]. Enchytraeids have a limited capacity for movement inside the soil; therefore, they live in the most superficial soil layers (0–10 cm) where the organic material and biological activity are concentrated. These organisms are found from arctic to tropical regions, and they are more abundant in forest soils (or soils rich in organic material) and less abundant in pastures and agricultural fields [52]. Their main food source is fungal mycelium; however, they also feed on organic matter that has been predigested by fungi as well as on other microorganisms [48].

These oligochaetes are sensitive to potentially toxic substances abundant in many soils where earthworms are not present or are not well represented. In addition, these animals are easy to handle and rear and have a significantly shorter life cycle than other worms, which is convenient for standardized toxicity assays [28, 43, 53]. These organisms live in close contact with the soil pore water and are exposed to soil contaminants through dermal, intestinal (through feeding), and respiratory routes [44]. Although their use in laboratory ecotoxicological tests was reported for the first time approximately 40 years ago, enchytraeids were selected for use in standardized ecotoxicological laboratory tests only 10 years ago as reported in the guidelines ISO 16387 [43] and OECD 220 [28].

Enchytraeus is the only enchytraeid genus with species selected for ecotoxicological tests standardized by ISO and OECD guidelines (e.g., *E. albidus* and *E. crypticus*) because this genus is considered typical of environmental stress indicator organisms and can be easily reared in the laboratory. *E. albidus* is the best known species for soil ecotoxicology [54] because it can be reared in many different substrates (it is widely distributed in terrestrial ecosystems) with different types of food, and it has a proven sensitivity to soil contaminants [55]. When these animals reach maturity (approximately 21 days at 18°C), the size of *E. albidus* adults can vary between 15 and 40 mm. Variations are observed in the total developmental cycle (33–74 days), embryonic development period (12–18 days), eggs per cocoon (7–10), and viable cocoon percent (40–50%) according to environmental conditions, especially in response to changes in culture temperature [43, 54]. The ideal reproduction temperature for the species is 15°C, although they can reproduce at temperatures between 12 and 22°C. Temperatures above 25°C should be avoided because they can suppress reproduction [54].

Although *E. crypticus* has lower ecological relevance because its prevalence in the field is unknown [56], the species is adequate for ecotoxicological laboratory tests, and its use in current standardized tests appears to be increasing [20]. Compared with *E. albidus*, *E. crypticus* has the advantage of being able to grow in agar medium, and it also has a higher reproduction rate, has a shorter generation time, tolerates a wider range of soil properties, and presents other characteristics considered methodologically advantageous [53, 57]. The adults of this species vary in sizes between 3 and 12 mm, and they have a generation time of approximately 18 days (at 21°C) in agar medium. The mean number of eggs per cocoon can vary between 1 and 35 with a mean of 4.6 eggs produced per day [58]. Studies have indicated that the number of juveniles may also vary according to the type of soil used, with apparently higher numbers in standard LUFA 2.2 natural soil [59] than in artificial OECD soil [60].

The order Collembola is one of the most diverse and abundant terrestrial arthropod orders, with 21 families and 20,000 described species [61]. In general, collembolans are small, varying from a few to approximately 10 mm in length [62]. The body can exhibit colorful pigmentation, although the inhabitants of deeper soil layers are typically not pigmented [63]. Most species feed on fungal hyphae and decomposing plant material; thus, they have a significant effect on microbial ecology and soil fertility and can control certain plant diseases caused by fungi [62, 64].

Collembolans are vulnerable to the presence of potentially toxic contaminants in soil [3] because they are exposed through water ingestion or absorption from wet/moist surfaces, food (living or dead) consumption, and soil pore air inhalation [44]. The responses obtained in tests using these arthropods may indicate environmental stress levels and the ecological risk of substances; therefore, these organisms supply information that can serve as a basis for legislation [65]. Collembolans have been used to estimate the effects of pesticides and other environmental pollutants on nontarget soil arthropods for approximately four decades, with *F. candida* – being typically used in standardized ecotoxicological tests [26, 31, 66]. *F. candida* has great importance for terrestrial ecosystem services because of its high sensitivity, short generation time, high reproduction rate, and easy culturing in the laboratory using a diet of granulated dry yeast [67].

F. candida (Willem 1902) is an arthropod of the family Isotomidae, and it is distributed in soils worldwide, although it is not common in most agricultural soils. This animal has a high occurrence rate in sites rich in organic matter [68]. This species has no pigmentation or eyes [38], and it reproduces exclusively through parthenogenetic females, which are approximately 2 mm long and sexually mature at 21–24 days of age (at 20°C). The optimal temperature for egg incubation and production is 21°C, and under these conditions, females lay 30–50 eggs, which take 7–10 days to hatch. This species has strong feeding preferences for certain species of fungi, and they are classified as microsaprophagous [62] and can be fed dry yeast in the laboratory [26, 31].

Although used in a fewer number of studies, *F. fimetaria* has been used as complementary or alternative species to *F. candida*, and the choice to use *F. fimetaria* is related to its higher ecological relevance because it is present in many natural and agricultural habitats where *F. candida* is not found. In addition, *F. fimetaria* meets all of the necessary requirements of a standard species for ecotoxicological laboratory tests [69]. In 2009, OECD guideline no. 232 [31] established *F. fimetaria* as a standard species for standardized laboratory assays. In addition to *F. fimetaria,* other species can be used to increase the ecological relevance of tests, although they may not be included in guidelines [21, 68, 70–71].

Mites are arthropods belonging to class Arachnida and subclass Acari, and they have a small size and unsegmented bodies [61]. In total, mites are ordered in 1,200 families and approximately 500,000 species [72], of which many are the most abundant mesofauna inhabitants in many types of soil and litter [73]. Suborder Gamasida (order Mesostigmata) includes the main species of predatory mites inhabiting soil pores [74]. The community structure and abundance of predatory mites are strongly dependent on the nature and availability of their prey [22]. Most mites feed on enchytraeids, nematodes, and microarthropods, although certain groups are considered fungivorous, bacteriophagous, facultative phytophagous, or have unknown feeding habits [73].

The ecosystem services provided by soil predatory mites include the biological control of pests and other species with abundant populations; thus, they significantly contribute to the flow of energy and matter in terrestrial ecosystems as well as to the maintenance of food chains [22]. In addition, several genera have been isolated for over 30 years from soil and tested as quality bioindicators [72]. Currently, reports are available for different toxicity tests using these arthropods; however, only the reproduction test using *H. aculeifer* has been standardized by guidelines for the evaluation of soil quality [20, 22, 30]. *H. aculeifer* has been considered the most adequate mite species for ecotoxicological assays because it has an acceptable generation time (approximately 1 month at 20°C), is a generalist predator, and can be easily handled in the laboratory [20, 72]. Because it represents a different trophic level from the other invertebrates used in standardized tests and is exposed to contaminants through different routes [44], *H. aculeifer* has been included on the EU community's list of nontarget organisms considered in the assessment of environmental risk of pesticides in soil [16, 72, 75].

Specimens of *H. aculeifer* are brown and have a light-brown dorsal shield. Although the size of both sexes is rather small, females are larger (0.8–0.9 mm) than males (0.55–0.65 mm) [22, 76]. Under temperatures between 20 and 23°C, these organisms become adults in approximately 16 (females) and 18 (males) days after going through larval, protonymph, and deutonymph developmental stages. However, their development time can be strongly affected by the temperature [76]. Usually, reproduction is sexual, although arrhenotokous parthenogenesis may occur in the absence of males, with this process only generating males [77]. However, females generally occur at a higher frequency because the sex ratio can also be controlled through selective cannibalism [78]. Each female lays approximately 100 eggs during its reproductive life [79]. These eggs are white, elliptical, and laid at the soil or culture substrate surface. Although it is a polyphagous predator, it is usually fed with the mites *Tyrophagus putrescentiae* or *Rhizoglyphus* sp. in ecotoxicological tests [30]. In the case of food scarcity, these organisms can survive cannibalistically [78].

In Ref. [22], the main invertebrates used in standardized terrestrial ecotoxicological tests were compared and it was concluded that the standard species of earthworms (*E. andrei* and *E. fetida*), enchytraeids (*E. albidus* and *E. crypticus*), collembolans (*F. candida* and *F. fimetaria*), and predatory mites (*H. aculeifer*) are adequate for the performance of tests on several soils from temperate and tropical regions. However, earthworms present limitations in acidic or basic soils. Enchytraeids are more tolerant to changes in soil pH and organic-matter concentrations, although they grow better in sandier soils. Mites and collembolans appear to be adequate for tests in most soil types and are considered less sensitive than oligochaetes (earthworms and enchytraeids) to soil properties. In tests that use soils with extreme characteristics (acidic or sandy soils), more than one species as well as alternative species should be used [22]. In addition, *H. aculeifer* generally appears to be less sensitive to certain substances compared with the remaining species. However, because it is the only predatory standard species, its inclusion in routine assessments of ecotoxicity is supported [80].

3. Summary of the standard procedures for bioassays with soil invertebrates

Performing ecotoxicological laboratory tests requires a series of steps, including planning and material preparation, animal rearing and maintenance in the laboratory, contamination of artificial/natural soils (or preparation of previously contaminated soil samples), experimental procedures for the initial test conditions as well as the maintenance and evaluation of toxicity tests using organisms, and data analysis and interpretation. The steps of the standardized tests using the species *E. andrei*/*E. fetida*, *E. albidus*/*E. crypticus*, *F. candida*, and *H. aculeifer* described in the guidelines ISO no. 11268-2 [27], no. 16387 [43], no. 11267 [26] and OECD no. 226 [30], respectively, will be summarized in this section to describe the main methods established by the guidelines and the adaptations that are used in regions with tropical climates.

Most standardized tests using soil invertebrates were developed to quantify the impact of chemical exposure on organisms in artificial soils [42]. However, although studies using artificial soils supply information that can be internationally compared, natural soils may provide information on local problems. Although the practice is still not described by the guidelines, the use of natural soils in standardized ecotoxicological tests has been increasing.

OECD soil is a standard artificial substrate recommended by ISO/OECD guidelines for most terrestrial ecotoxicology studies [42]. This substrate consists of a mixture of 70% industrial sand (with more than 50% particles between 0.05 and 0.2 mm), 20% kaolinite clay, and 10% peat (ground and dry). However, for assays in tropical regions, studies have used a modified version of this substrate [81–83] known as tropical artificial soil (TAS). TAS uses powdered coconut husks as replacement for peat because of its higher availability in tropical regions. In both cases, after the materials are mixed, the pH (1 M KCL 1:5 weight:volume ratio) of the artificial soil should be adjusted to 6.0 ± 0.5 through the addition of $CaCO_3$. In addition, the soil water-holding capacity should be determined for moisture adjustments [26].

To use natural soils in standardized tests, the soils should offer minimum conditions for the survival and reproduction of the test species without causing morphological or behavioral changes in the absence of the contaminant. To test chemicals for regulation purposes (e.g., ecological risk of pesticides), natural soils are artificially contaminated, and the results should be comparable between laboratories. Therefore, the soils should have similar characteristics, as is the case for LUFA soils [84–85], EURO soils [86], SIM soils [87], and other natural soils selected as standard soils for specific regions, for example, Polish [88] and Mediterranean soils [89]. In the case of natural soils from contaminated areas, the use of a control soil with the same characteristics (texture, pH, organic matter concentration, and C:N ratio) but without contaminants is recommended [87].

Natural soils should be dried, sieved (5 or 2 mm, preferentially), and chemically and physically characterized before use, and at least the texture, pH, WHC, and moisture content should be determined. In addition, the organic matter concentration (or organic C), CEC, C:N ratio, and metal (or other element) concentration may also be measured [87]. Regardless of the type of soil (artificial or natural), defauning is recommended (by soil freezing at −20°C followed by thawing to room temperature) to eliminate the original soil fauna organisms [90].

According to the guidelines, incorporating test substances directly into soil (artificial contamination) varies with the water solubility of the contaminant, and there are three main methods: (a) for water-soluble substances; (b) for water-insoluble but organic solvent-soluble substances; and (c) for water-insoluble and organic solvent-insoluble substances. In all cases, the concentration gradients of the test substances should be prepared immediately before the beginning of the assay in the volume necessary to maintain the soil moisture between 40 and 60% of its WHC. In the case of water-insoluble substances, after applying the solutions in increasing concentrations, the soil moisture should be adjusted through the addition of pure water. It is recommended that the concentrations be prepared in a geometric series separated by a factor of 1.8 or lower. If effects are not observed for the tested substance (e.g., active ingredient of a pesticide) at the highest concentration (1,000 mg kg^{-1}) in the preliminary tests (acute toxicity assays), then a limit test should be performed to evaluate the toxicity using only the control treatment at a concentration of 1,000 mg kg^{-1}.

The environmental conditions of the standardized ecotoxicological tests and laboratory cultures should be controlled. Controlled temperatures and light conditions in the test/culture chamber/room are fundamental for obtaining homogeneous cultures (with the same age and size) with development cycles that occur within the time predicted for the tests. The main recommended protocols are a mean temperature of 20 ± 2°C and constant light intensity between 400 and 800 lux on the culture containers [26, 31]. In addition, it is recommended that cultures be kept under controlled light/dark cycles, preferably 12 hours light/12 hours dark or 16 hours light/8 hours dark. However, the environmental test and culture conditions have been adapted for regions with predominantly tropical climates to better reflect the influence of local climate conditions. To simulate tropical conditions, temperatures varying between 23 and 27°C have been used [83, 91–93].

The earthworms *E. andrei* and *E. fetida* should be reared in substrates composed of a mix of horse or cow manure (defauned following the same process described for soils) and peat (1:1, dry weight) [27], and the culture medium should be moistened weekly with pure water. Similar to artificial soils, peat is usually replaced with powdered coconut husks in tropical climate regions [13, 18, 81]. Earthworms should be fed weekly with a mixture of oat flakes and water, and the rearing substrate should be periodically replaced.

E. crypticus can be cultured in natural soil, although recent studies have opted for culturing in Petri dishes with agar medium [83, 94]. This culture medium is composed of a mixture (1:1, v:v) of a salt solution (calcium chloride, magnesium sulfate, potassium chloride, and sodium bicarbonate) and a bacto-agar solution (e.g., Oxoid - Agar No. 1). Approximately 50 mg of ground oat flakes should be supplied as food once a week, and the organisms should be transferred to new substrate every 2 months.

Laboratory cultures of collembolans and mites should be performed using a substrate composed of a mixture of activated charcoal, plaster of Paris (calcium sulfate), and deionized water, with a recommended 1:8 charcoal:plaster of Paris ratio for *F. candida* and 1:9 ratio for *H. aculeifer* (w:w) [26, 30]. The volume of deionized water should be 60–100 mL for each 100 g of mixture, although the water content varies with the type of plaster of Paris. The bottom of plastic containers should be filled with the mixture to a height of approximately 1 cm. Collembolans should be fed dry yeast once a week, and cultures of *T. putrescentiae* or *Caloglyphus* sp. (cheese mites) should be simultaneously maintained with *H. aculeifer* cultures to serve as food (prey) for the predator. Small quantities of cheese mites should be supplied to *H. aculeifer* twice a week and the cheese mites should be fed once a week with powdered brewer's yeast [30].

Avoidance assays with *E. andrei*/*E. fetida* or *F. candida,* which are described by the guidelines ISO:17512-1 and ISO:17512-2 [41, 95], are performed using rectangular (earthworms) or round (collembolans) plastic boxes that are divided into two equal compartments by a plastic divider vertically introduced and filled with the test soil (amounts depend on the size of the container). Contaminated soil is added to one of the compartments, and the same amount of the respective control soil is added to the other compartment. Ten *E. andrei* or *E. fetida* adults (with developed clitellum) or 20 *F. candida* individuals at 10–12 days of age (originating from synchronized cultures) are then placed on the separation line between the two compartments, and the plastic divider is removed. The containers are then closed with perforated lids to allow air circulation. The animals are not fed during the test. After 48 hours, the number of individuals present in each compartment is recorded according to the specific method for each species [18, 96]. A double control test is also performed with control soil in both compartments to determine whether the organisms are randomly distributed between the two compartments in the absence of contaminants.

Chronic toxicity tests with *E. andrei*/*E. fetida* should be performed according to ISO:11268-2 [27]. Round plastic containers are filled with approximately 500 g soil (dry weight) treated with solutions with increasing contaminant concentrations (or dilutions of the naturally contaminated soils) or the respective control soils. Ten adult earthworms (with visible clitellum) with individual weights between 300 and 600 mg that had been previously incubated in control soil for at least 24 hours are selected for each experimental unit. The containers are closed with perforated lids, and the earthworms are fed horse manure (≈5 g per replicate) at the beginning of the assay and then once a week until the end of the assay. The assay lasts for a total of 56 days. Adult earthworms are removed after 28 days, washed with water, and weighed. The final percentage of body biomass (after 28 days) relative to the initial weight can be calculated to assess the effects of contaminants on organismal growth. After 56 days from the beginning of the assay, the number of juveniles is counted to verify the treatment effect on species reproduction (for additional details, see Ref. [27], and/or Ref. [18]).

Reproduction tests with *E. crypticus* are described in ISO:16387 [43]. Ten similar-sized adults (with visible clitellum) are placed in cylindrical containers containing 20 g soil (dry weight) that had been treated with the test substance or the respective control soil. Finely ground oat flakes can be supplied as food (≈2 mg per replicate), and the containers should be hermetically closed [83]. The containers are opened weekly to allow for gas exchange, and food and soil moisture are replenished as needed. Twenty-eight days following the beginning of the assay, the total number of enchytraeids are counted using a stereoscopic microscope following fixation in 80% ethanol, staining with rose bengal (1% in ethanol), and wet sieving of the organisms [94]. For *E. albidus*, the experimental procedures are somewhat different, especially with regard to the assay duration.

Tests evaluating the impact of contaminants on the reproduction of the collembolan *F. candida* are described in ISO:11267 [26]. Thirty grams of contaminated or control soil (fresh weight) are added to cylindrical containers (approximately 100 mL). Ten adult collembolans aged between 10 and 12 days (originating from synchronized cultures) are then placed into each experimental unit, and the containers are then hermetically closed. Food (dry yeast) is supplied at the beginning of the assay and on the 14th day, and the containers are opened weekly to allow for gas exchange. On the 28th day after the beginning of the assay, the soil of each replicate is submersed in water to force the survivors to float to the surface, and the juveniles are counted following the addition of several drops of black ink (for increased contrast) [93].

Reproduction tests with *H. aculeifer* are described in guideline OECD no. 226 [30]. The containers are filled with 20 g (fresh weight) of soil treated with the test substance or control soil. Ten females with ages between 28 and 35 days and originating from the synchronized cultures are then placed in each container. The animals receive small amounts of food (cheese mites) at the beginning of the assay and then twice a week until the end of the assay. The containers are then hermetically closed and opened weekly for airing and soil moisture adjustments. Fourteen days after the beginning of the assay, the mites are removed from the soil using a MacFadyen extractor with a gradient of increasing temperatures for 48 hours (12 hours at 25°C, 12 hours at 35°C, and 24 hours at 45°C). Adults and newly emerged juveniles in each replicate should be fixed in 70% ethanol and counted using a stereoscopic microscope.

The following parameters are used to evaluate the critical values: NOEC (no observed effect concentration) and LOEC (lowest observed effect concentration), LC_{10} and LC_{50} (lethal concentration to 10 and 50% of the population, respectively), EC_{10} and EC_{50} (concentration at which 10 and 50% of the contaminant's maximal effect is observed on the growth or number of juveniles, respectively), and AC_{50} (concentration causing the avoidance of 50% of the organisms from the contaminated soils). These toxicity parameters are designed to detect the ecological risk of the chemicals in soil and used to derive protection limits for terrestrial ecosystems.

The significance of avoidance responses (LOEC and NOEC) is tested with Fisher's exact test using a two-tailed test for the double control conditions and a one-tailed test for the contaminated soil combination conditions [97]. The AC_{50} values can be obtained using regression analyses [98]. The significance of the effects (LOEC and NOEC) on the body biomass of the earthworms and number of earthworms, enchytraeids, mites, and collembolan juveniles following exposure to the contaminated soils should be tested using a one-way analysis of variance (ANOVA). When differences are detected ($p \leq 0.05$), the treatment means should be compared with the results from the respective controls using a post hoc test such as Dunnett's test. The EC_{10} and EC_{50} values should be estimated by nonlinear regression using pre-defined exponential, logistic, Gompertz, or hormesis models [71]. The normality and homogeneity of variance should be tested prior to the ANOVAs.

4. Future prospects in soil ecotoxicology

The future prospects for soil ecotoxicology refer to the challenge of increasing the realism of the analyses in the terrestrial environment and to reduce the uncertainties about the real degree of ecological risk obtained by laboratory tests. To improve the ecological relevance of laboratory assays, it needs a transition from research based on artificial soil to the use of natural soils, which consider the real relationship between contaminants and test organisms in the exposure scenarios. Moreover, it is necessary to increase the list of standard organisms for the tests, considering the practicality in assays, and especially the inclusion of species that represent the geographical and ecological conditions of the test site [99]. As an example, Ref. [20] suggests the standardization of the sublethal toxicity tests with isopods, since they represent an ecologically relevant group of soil fauna, and the effects on these arthropods can be determined at biochemical, genomic, individual (growth, behavior), and ecological (feeding activity) levels.

It is still necessary to move forward in the assessments of long-term sublethal effects; besides, there is a need for better understanding of exposure, absorption, and metabolism of substances in individuals, and the identification of the responses at different levels of biological organization (e.g., communities) [20]. Based on this assumption, one of the apparent possibilities to evaluate the long-term impact in standard laboratory tests would be the use of multigenerational assays, where toxic effects such as delayed reproductive failures, transmission of the bioaccumulation to offspring, or accumulation of DNA damage could be identified [100]. For a better understanding of the relationships between pollutants and species, more studies using chemical, biochemical, and molecular prospects (ecotoxicogenomics) are needed, particularly assessments of bioavailability, bioaccumulation, and molecular biomarkers [20, 101]. The assessment of impacts at higher levels of biological organization can be accomplished through multispecies assays, which consider the relationships between species, contaminants, and soil properties. In addition, the semi-field and field tests may offer a better understanding of the contaminant's impacts on soil communities, although they are more complex, especially when there is the involvement of comparisons between different ecosystems [13, 20].

Finally, it is necessary that future adjustments be performed in the standard assays available, in order to enable them to address the new and emerging needs of the current ecotoxicology, such as the case of evaluation of the toxicity of nanoparticles and mixtures of contaminants, among others [20]. These adjustments also extend to the studies performed in tropical regions, where there is a need for a revision of the methods, especially in terms of soils, species, and climatic conditions, in order to increase the ecological relevance of the analyses at local level.

Author details

Paulo Roger Lopes Alves[1*] and Elke Jurandy Bran Nogueira Cardoso[2]

*Address all correspondence to: paulo.alves@uffs.edu.br

1 Federal University of South Border, Chapecó, Brazil

2 Luiz de Queiroz College of Agriculture -University of São Paulo, Piracicaba, Brazil

References

[1] Decaëns T, Jimenez JJ, Gioia C, Measey GJ, Lavelle P. The values of soil animals for conservation biology. European Journal of Soil Biology. 2006;S23–S38.

[2] Pey B, Nahmani J, Auclerc A, Capowiez Y, Cluzeau D, Cortet J, Decaëns T, Deharveng L, Dubs F, Joimel S, Briard C, Grumiaux F, Laporte M, Pasquet A, Pelosi C, Pernin C, Ponge J, Salmon S, Santorufo L, Hedde M. Current use of and future needs for soil invertebrate functional traits in community ecology. Basic and Applied Ecology. 2014;15(3):194–206.

[3] Pankhurst CE, Doube BM, Gupta VVSR. Biological indicators of soil health: Synthesis. In: Pankhurst CE, Doube BM, Gupta VVSR, editors. Biological Indicators of Soil Health. Wallingford: CAB International; 1997. pp. 265–296.

[4] Six J, Feller C, Denef K, Ogle SM, Sa JCD, Albrecht A. Soil organic matter, biota and aggregation in temperate and tropical soils – effects of no-tillage. Agronomie. 2002;22:755–775.

[5] Lavelle P, Decaëns T, Aubert M, Barot S, Blouin M, Bureau F, Margerie P. Soil invertebrates and ecosystem services. European Journal of Soil Biology. 2006;42:S3–S15.

[6] Cortet J, Gomot-de-Vauflery A, Poinsot-Balaguer N, Gomot L, Texier C, Cluzeau D. The use of invertebrate soil fauna in monitoring pollutant effects. European Journal of Soil Biology. 1999;35(3):115–134.

[7] Moser H, Römbke J, editors. Ecotoxicological Characterization of Waste – Results and Experiences of an International Ring Test. Chichester: Wiley. 2009;p. 308 p.

[8] Parisi V, Menta C, Gardi C, Jacomini C, Mozzanica E. Microarthropod communities as a tool to assess soil quality and biodiversity: A new approach in Italy. Agriculture, Ecosystems & Environment. 2005;105:323–333.

[9] Cardoso EJBN, Vasconcellos LR, Bini D, Yumi MM, Santos CA, Alves PRL, de Paula AM, Nakatani AS, Pereira JM, Nogueira MA. Soil health: Looking for suitable indicators. What should be considered to assess the effects of use and management on soil health? Scientia Agricola. 2013;70:274–289.

[10] Schoenholtz SH, Vam Miegroet H, Burger JA. A review of chemical and physical properties as indicators of forest soil quality: Challenges and opportunities. Forest Ecology and Management. 2000;138:335–356.

[11] Everts JW, Aukema B, Hengeveld R, Koeman JH. Side-effects of pesticides on ground-dwelling predatory arthropods in arable ecosystems. Environmental Pollution. 1989;59:203–225.

[12] Krogh PH. Microarthropods as bioindicators [thesis]. Arhus: University of Arhus; 1994.

[13] Cardoso EJBN, Alves PRL. Soil ecotoxicology. In: Begum G, editor. Ecotoxicology. Rijeka: InTech – Open Access Publisher. 2012; pp. 27–50.

[14] Melo LCA. Phytotoxicity of cadmium and barium and derivation of critical limits in soils [thesis]. Piracicaba: University of São Paulo; 2010.

[15] Shugart LR, editor. Emerging Topics in Ecotoxicology: Principles, Approaches and Perspectives. Oak Ridge: Springer Science; 2009. 400p.

[16] EC. 284/2013 of 1 March 2013 Setting Out the Data Requirements for Plant Protection Products, in Accordance with Regulation (EC) No 1107/2009 of the European Parliament and of the Council Concerning the Placing of Plant Protection Products on the Market. J.O.L. 2013;93:1–84.

[17] EC. Directive 2008/98/EC: Directive of the European Parliament and of the Council of 19 November 2008 on Waste and Repealing Certain Directives. Official Journal of the European Union. 2008;312:3–30.

[18] Alves PRL, Cardoso EJBN, Martines AM, Sousa JP, Pasini A. Earthworm ecotoxicological assessments of pesticides used to treat seeds under tropical conditions. Chemosphere. 2013;90:2674–2682.

[19] Méhu J, Bazin C, Grelier-Volatier L. Classification réglementaire et écocompatibilité des déchets. Techniques de l'ingénieur. 2004;G2-030:1–10.

[20] Van Gestel CAM. Soil ecotoxicology: State of the art and future directions. ZooKeys. 2012;176:275–296.

[21] Kapanen A, Itavaara M. Ecotoxicity tests for compost applications. Ecotoxicology and Environmental Safety. 2001;49:1–16.

[22] Jänsch S, Amorim MJ, Römbke J. Identification of the ecological requirements of important terrestrial ecotoxicological test species. Environmental Review. 2005;13:51–83.

[23] Kula H, Larink O. Development and standardization of test methods for the prediction of sublethal effects of chemicals on earthworms. Soil Biology and Biochemistry. 1997;29:635–639.

[24] Rida AMMA, Bouché MB. Earthworm toxicology: From acute to chronic tests. Soil Biology and Biochemistry 1997;29:699–703.

[25] Hoffman DJ, Rattner BA, Burton GA, Cairns J, editors. Handbook of Ecotoxicology. Dordrecht. v. 2. London: Blackwell; 2003. 1290 p.

[26] ISO. Soil Quality no.11267: Inhibition of Reproduction of Collembola (*Folsomia candida*) by Soil Pollutants. Geneva: International Standardization Organization; 1999.

[27] ISO. Soil quality no. 11268-2: Effects of pollutants on earthworms – Part 2: Determination of effects on reproduction of *Eisenia fetida/Eisenia andrei*. Geneva: International Standardization Organization; 2012.

[28] OECD. Guidelines for the Testing of Chemicals No. 220: Enchytraeid Reproduction Test. Paris: Organization for Economic Co-operation and Development; 2004a.

[29] OECD. Guidelines for the Testing of Chemicals No. 222: Earthworm Reproduction Test (*Eisenia fetida/Eisenia andrei*). Organization for Economic Co-operation and Development; 2004b.

[30] OECD. Guidelines for the testing of Chemicals No. 226: Predatory Mite [*Hypoaspis (Geolaelaps) aculeifer*] Reproduction Test in Soil. Organization for Economic Co-operation and Development; 2008.

[31] OECD. Guidelines for the testing of Chemicals No. 232: Collembolan Reproduction Test in Soil. Organization for Economic Co-operation and Development; 2009.

[32] Yearcley RB, Lazorchak JM, Gast LC. The potential of an earthworm avoidance test for evaluation of hazardous waste sites. Environmental Toxicology and Chemistry. 1996;15:1532–1537.

[33] Hund-Rinke K, Achazi R, Römbke J, Warnecke D. Avoidance test with *Eisenia fetida* as indicator for the habitat function of soils: Results of a laboratory comparison test. Journal of Soils and Sediments. 2003;3(1)7–12.

[34] Edwards CA, Bohlen PJ, editors. Biology and Ecology of Earthworms. London: Chapman and Hall; 1995. 436 p.

[35] Heupel K. Avoidance response of different collembolan species to Betanal. European Journal of Soil Biology. 2002;38:273–276.

[36] Langdon CJ, Hodson ME, Arnold RE, Black S. Survival, Pb-uptake and behavior of three species of earthworm in Pb treated soils determined using an OECD-style toxicity test and a soil avoidance test. Environmental Pollution. 2005;138:368–375.

[37] Ronday R, Houx NWH. Suitability of seven species of soil-inhabiting invertebrates for testing toxicity of pesticides in soil pore water. Pedobiologia. 1996;40:106–112.

[38] Fountain MT, Hopkin SP. *Folsomia candida* (Collembola): A "Standard" soil arthropod. Annual Review of Entomology. 2005;50:201–22.

[39] Van Gestel CAM, Léon CD, Van Straalen NM. Evaluation of soil fauna ecotoxicity tests regarding their use in risk assessment. In: Tarradellas J, Bitton G, Rossel D, editors. Soil Ecotoxicology. Boca Raton: CRC Press; 1997. pp. 291–317.

[40] ISO. Soil quality No. 11268-1: Effects of pollutants on earthworms (*Eisenia fetida*) – Part 1: Determination of acute toxicity using artificial soil substrate. Geneva: International Standardization Organization; 1993.

[41] ISO. Soil quality No. 17512-1: Avoidance Test for Determining the Quality of Soils and Effects of Chemicals on Behavior – Part 1: Test with Earthworms (*Eisenia fetida* and *Eisenia andrei*). Geneva: International Standardization Organization; 2008.

[42] OECD. Guidelines for the Testing of Chemicals No. 207: Earthworm, Acute Toxicity Tests. Paris: Organization for Economic Co-operation and Development; 1984.

[43] ISO. Soil quality No. 16387: Effects of Pollutants on Enchytraeidae (*Enchytraeus sp.*) – Determination of Effects on Reproduction and Survival. Geneva: International Standardization Organization; 2004.

[44] Peijnenburg W, Capri E, Kula C, Liess M, Luttik R, Montforts M, Nienstedt K, Rombke J, Sousa JP, Jensen J. Evaluation of exposure metrics for effect assessment of soil invertebrates. Critical Reviews in Environmental Science and Technology. 2012;42:1862–1893.

[45] Edwards CA, editor. Earthworm Ecology. V. 3. Boca Raton: CRC Press LLC; 2004. 438 p.

[46] Domínguez J, Ferreiro A, Velando A. Are *Eisenia fetida* (Savigny, 1826) and *Eisenia andrei* Bouché, 1972 (Oligochaeta, Lumbricidae) different biological species? Pedobiologia. 2005;49:81–87.

[47] Hassal L, Turner JG, Rands MRW. Effects of terrestrial isopods on the decomposition of woodland leaf litter. Oecologia. 1987;72:23–36.

[48] Didden WAM. Ecology of terrestrial Enchytraeidae. Pedobiologia. 1993;37:2–29

[49] Dash MC. Oligochaeta: Enchytraeidae. In: Dindal DL, editor. Soil Biology Guide. New York: John Wiley & Sons; 1990. pp. 311–40.

[50] Laakso J, Setälä H. Sensitivity of primary production to changes in the architecture of belowground food webs. Oikos. 1999;87:57–64.

[51] Van Vliet PCJ. Enchytraeids. In: Sumner M, editor. The Handbook of Soil Science. Boca Raton: CRC Press; 2000. pp. C70–C75.

[52] ADL SM, editor. The Ecology of Soil Decomposition. London: CAB International; 2003. 350 p.

[53] Castro-Ferreira MP, Roelofs D, Van Gestel CAM, Verweij RA, Soares AMVM, Amorim MJB. *Enchytraeus crypticus* as model species in soil ecotoxicology. Chemosphere. 2012;87(11):1222–1227.

[54] Römbke J, Moser T. Validating the enchytraeid reproduction test: organization and results of an international ringtest. Chemosphere. 2002;46:1117–1140.

[55] Jeffery SS, Gardi C, Jones A, Montanarella L, Marmo L, Miko L, Ritz K, Pérès G, Römbke J, van der Putten W, editors. European Atlas of Soil Biodiversity. Luxemburg: European Commission – Publication Office of the European Union; 2010.

[56] Römbke J. Ecotoxicological laboratory tests with enchytraeids: A review. Pedobiologia. 2003;47:607–616.

[57] Van Gestel CAM, Borgman E, Verweij RA, Diez OM. The influence of soil properties on the toxicity of molybdenum to three species of soil invertebrates. Ecotoxicology and Environmental Safety. 2011;74:1–9.

[58] Westheide W, Graefe U. Two new terrestrial Enchytraeus species (Oligochaeta, Annelida). Journal of Natural History. 1992;26:479–488.

[59] Achazi RK, Chroszcz G, Mierke W. Standardization of test methods with terrestrial invertebrates for assessing remediation procedures for contaminated soils. Eco-Informa. 1997;12:284–89.

[60] Kuperman RG, Amorim MJB, Römbke J, Lanno R, Checkai RT, Dodard SG, Sunahara GI, Scheffczyk A. Adaptation of the enchytraeid toxicity test for use with natural soil types. European Journal of Soil Biology. 2006;42:S234–S243.

[61] Lavelle P, Spain AV, editors. Soil Ecology. New York: Kluwer Academic Publishers; 2003. 682 p.

[62] Hopkin SP. Biology of springtails. Insecta: Collembola. Oxford: Oxford University Press; 1997. 330 p.

[63] Petersen H. Collembolan ecology at the turn of the millennium. Pedobiologia. 2002;46:246–260.

[64] Culik, MP, Zeppelini D. Diversity and distribution of Collembola (Arthropoda: Hexapoda) of Brazil. Biodiversity and Conservation. 2003;12:1119–1143.

[65] Van Straalen NM. Assessment of soil contamination – a functional perspective. Biodegradation. 2002;13:41–52.

[66] Krogh PH. Toxicity Testing with the Collembolans *Folsomia fimetaria* and *Folsomia candida* and the Results of a Ring Test (trans: DMU/AU DoTE). Miljoministeriet; 2008. 44 p.

[67] Crouau Y, Gisclard C, Perotti P. The use of *Folsomia candida* (Collembola, Isotomidae) in bioassays of waste. Applied Soil Ecology. 2002;19:65–70.

[68] Wiles JA, Krogh PH. Testing with the collembolans *I. viridis, F. candida* and *F. fimetaria*. In: Løkke H, Van Gestel CAM, editors. Handbook of Soil Invertebrate Toxicity Tests. Chichester: John Wiley & Sons; 1998. pp. 131–156.

[69] Van Gestel CAM. The developmental status of ecotoxicity tests. Løkke H, Van Gestel CAM, editors. Handbook of Soil Invertebrate Toxicity Tests. Chichester: John Wiley & Sons; 1998.pp. 57–67.

[70] Greenslade P, Vaughan GT. A comparison of Collembola species for toxicity testing of Australian soils. Pedobiologia. 2003;47:171–179.

[71] Environment-Canada. Guidance Document on Statistical Methods for Environmental Toxicity Tests. Environmental Protection Series, EPS 1/RM/46, 2005 with 2007 updates. Ottawa: Environment Canada; 2007.

[72] Huguier P, Manier N, Owojori OJ, Bauda P, Pandard P, Römbke J. The use of soil mites in ecotoxicology: A review. Ecotoxicology. 2015;24(1):1–18.

[73] Mineiro JC, Moraes GJ. Gamasida (Arachnida: Acari) Edáficos de Piracicaba, Estado de São Paulo. Neotropical Entomology. 2001;30(3):379–385.

[74] RömbkeJ, Beck L, Dreher P, Hund-Rinke K, Jänsch S, Kratz W, Pieper S, Ruf A, Spelda J, Woas S. Entwicklung von bodenbiologischen Bodengüteklassen für Acker- und Grünlandstandorte. UBA-Texte 20/02; 2002. 273 p.

[75] Smit CE, Moser T, Rombke J. A new OECD test guideline for the predatory soil mite *Hypoaspis aculeifer*: Results of an international ring test. Ecotoxicology and Environmental Safety. 2012;82:56–62.

[76] Helyer N, Brown K, Cattlin ND. A Colour Handbook of Biological Control in Plant Protection. London: Manson Publishing; 2003. 128 p.

[77] Krogh PH, Axelsen JA. Test on the predatory mite *Hypoaspis aculeifer* preying on the collembolan *Folsomia fimetaria*. In: Løkke H, Van Gestel CAM, editors. Handbook of Soil Invertebrate Toxicity Tests. Chichester: John Wiley & Sons; 1998. pp. 239–251.

[78] Berndt O, Meyhöfer R, Poehling HM. Propensity towards cannibalism among *Hypoaspis aculeifer* and *H. miles*, two soil-dwelling predatory mite species. Experimental and Applied Acarology. 2003;31(1–2):1–14.

[79] Ruf A. Die morphologische Variabilität und Fortpflanzungsbiologie der Raubmilbe *Hypoaspis aculeifer* (Canestrini 1883) (Mesostigmata: Laelapidae) [thesis]. Bremen: University of Bremen; 1993.

[80] Owojori OJ, Waszak K, Römbke J. Avoidance and reproduction tests with the predatory mite *Hypoaspis aculeifer*: Effects of different chemical substances. Environmental Toxicology and Chemistry. 2014;33:230–237.

[81] Garcia MV. Effects of pesticides on soil fauna: Development of ecotoxicological test methods for tropical regions [thesis]. Zentrum für Entwicklungsforschung: University of Bonn; 2004.

[82] Römbke J, Garcia MVB, Scheffczyk A. Effects of the fungicide benomyl on earthworms in laboratory tests under tropical and temperate conditions. Archives of Environmental Contamination and Toxicology. 2007;53:590–598.

[83] Alves PRL, Natal-Da-Luz T, Sousa JP, Cardoso EJBN. Ecotoxicological characterization of sugarcane vinasses when applied to tropical soils. Science of the Total Environment. 2015;526:222–232.

[84] Schinkel K. Prüfung der Beständigkeit von Pflanzenschutzmitteln im Boden im Rahmen des Zulassungsverfahrens. Ber Landwirtsch. 1985;198:9–20.

[85] Lokke H, Gestel CAM, editors. Handbook of Soil Invertebrate Toxicity Tests. Chichester: Wiley & Sons, 1985. 281 p.

[86] Kuhnt G, Vetter L. Field work of EUROSOILS: Sampling sites, profile properties and collection procedures. In: Gawlik BM, Muntau H, editors. EUROSOILS II—Laboratory Reference Materials for Soil-related Studies. Luxembourg: European Commission; 1999. pp. 41–56.

[87] Römbke J, Amorim M. Tackling the heterogeneity of soils in ecotoxicological testing: An EURO-soil based approach. Journal of Soils and Sediments. 2004;4:276–281.

[88] Maliszewska-Kordybach B, Klimkowicz-Pawlas A, Smreczak B. Soil reference materials in ecotoxicity testing—application of the concept of EURO-soils to soils from Poland. Polish Journal of Environmental Studies. 2008;17:257–266.

[89] Chelinho S, Domene X, Campana P, Natal-da-Luz T, Scheffczyk A, Römbke J, Andrés P, Sousa JP. Improving ecological risk assessment in the Mediterranean area: Selection of reference soils and evaluating the influence of soil properties on avoidance and reproduction of two oligochaete species. Environmental Toxicology and Chemistry. 2011;30(5):1050–8. doi:10.1002/etc.480

[90] Pesaro M, Widmer F, Nicollier G, Zeyer J. Effects of freeze-thaw stress during soil storage on microbial communities and methidathion degradation. Soil Biology and Biochemistry. 2003;35:1049–1061.

[91] Niemeyer JC, Moreira-Santos M, Ribeiro R, Da Silva EM, Sousa JP. Environmental risk assessment of a metal contaminated area in the tropics. Tier I: Screening phase. Journal of Soils and Sediments. 2010;10:1557–1571.

[92] Natal-Da-Luz T, Moreira-Santos M, Ruepert C, Castillo LE, Ribeiro R, Sousa JP. Ecotoxicological characterization of a tropical soil after diazinon spraying. Ecotoxicology. 2012;21:2163–2176.

[93] Alves PRL, Cardoso EJBN, Martines AM, Sousa JP, Pasini A. Seed dressing pesticides on springtails in two ecotoxicological laboratory tests. Ecotoxicology and Environmental Safety. 2014;105:65–71.

[94] Chelinho S, Domene X, Campana P, Andrés P, Römbke J, Sousa JP. Toxicity of phenmedipham and carbendazim to *Enchytraeus crypticus* and *Eisenia andrei* (Oligochaeta) in Mediterranean soils. Journal of Soils and Sediments. 2014;14:584–599.

[95] ISO. Soil quality No. 17512-2: Avoidance Test for Determining the Quality of Soils and Effects of Chemicals on Behaviour – Part 2: Test with Collembolans (*Folsomia candida*). Geneva: International Standardization Organization; 2011.

[96] Natal-Da-Luz T, Tidona S, Jesus B, Morais PV, Sousa JP. The use of sewage sludge as soil amendment. The need for an ecotoxicological evaluation. Journal of Soils and Sediments. 2009;9(3):246–260.

[97] Zar JH. Biostatistical Analysis. London: Prentice Hall International; 1999. 663 p.

[98] Sakuma M. Probit analysis of preference data. Applied Entomology and Zoology. 1998;33:339–347.

[99] Kuperman RG, Checkai RT, Vinicius M, Garcia B, Römbke J. State of the science and the way forward for the ecotoxicological assessment of contaminated land. Pesquisa Agropecuária Brasileira. 2009;1:811–824.

[100] Paumen ML, Steenbergen E, Kraak MHS, Van Straalen NM, Van Gestel, CAM. Multigeneration exposure of the springtail *Folsomia candida* to phenanthrene: From dose-response relationships to threshold concentrations. Environmental Science and Technology. 2008;42:6985–6990.

[101] Filser J, Wiegmann S, Schröder B. Collembola in ecotoxicology — Any news or just boring routine? Applied Soil Ecology. 2014;83:193–199.

Chapter 4

Rotifers as Models in Toxicity Screening of Chemicals and Environmental Samples

Roberto Rico-Martínez, Mario Alberto Arzate-Cárdenas, Daniel Robles-Vargas, Ignacio Alejandro Pérez-Legaspi, Alvarado-Flores Jesús and Gustavo Emilio Santos-Medrano

Additional information is available at the end of the chapter

http://dx.doi.org/10.5772/61771

Abstract

An important objective of aquatic ecotoxicology is to determine the effects of toxic compounds in organisms that play a central role in aquatic communities where rotifers have a large impact on several important ecological processes. The contribution of the rotifers to secondary production in many aquatic communities is substantial as they are often the larger fraction of zooplankton biomass at certain times of the year. In addition to the importance of their ecological roles in aquatic communities, the rotifers are attractive organisms for ecotoxicological studies by its short life cycles and rapid reproduction, their small size, and little volumes needed for culture and toxicity assays. The main end points used in ecotoxicological studies are mortality, reproduction, behavior, and biomarkers. Such parameters are included in international regulations from all over the world, where different species are used to evaluate the effect of environmental samples or chemical compounds. The high diversity of rotifers is an important issue because it can modify their relative susceptibility to toxicants. Thus, more studies are needed to know the relations and mechanisms involved in clonal variation, sensitivity, and development, which can be all assessed by state-of-the-art procedures.

Keywords: aquatic toxicology, ecotoxicology, metal toxicity, acute toxicity, endocrine disruption

1. Introduction

The analytical equipment can identify and quantify a chemical substance but not its toxicity in the organisms or the environment, which can be evaluated only in life organisms [1]. Toxicity testing in water samples assesses the concentration and exposure time of the chemical substances that produce an adverse effect in aquatic organisms, generating useful data for risk assessment. A toxicity test can be accepted by the scientific community if (a) it is capable to predict adverse effects for a variety of compounds in different organisms, (b) it must be replicable with statistical-based analysis, (c) its data must include adverse effects in a range of concentrations in real exposure times, (d) it must be useful to evaluate a risk, (e) it is economic and easy to perform, and (f) it is sensitive and realistic [2].

A model organism for a toxicity test must be abundant, native, or representative of the ecosystem that would be impacted, with ecological and economic importance; there must be a good knowledge of its basic biology that helps interpretation of data and be available for routine maintenance in the laboratory [2].

Toxicity tests have been developed for rotifers to assess many and diverse end points like (a) mortality, (b) reproduction, (c) production of amictic females, (d) cyst production, (e) probability of extinction, (f) behavior, (g) ingestion rates, (h) swimming activity, (i) *in vivo* enzymatic activity, and (j) genetic expression, among others [3–5].

Most of the toxicity tests in rotifers are lethal tests that rely in mortality as the end point to calculate the LC50 value typically at 24 or 48 h without feeding [5]. These tests once standardized are the base to develop monitoring protocols to assess water quality [4, 5]. On the other hand, chronic toxicity tests assess sublethal parameters like behavioral, physiological, or reproductive alterations as the first responses to toxic substances [5, 6], which show high sensitivity in shorter periods of time. The rotifer species *Brachionus calyciflorus* and *Brachionus plicatilis* are among the species most frequently used for both lethal and chronic tests [7, 8].

1.1. Phylum Rotifera

Rotifers are aquatic or semiaquatic microscopic invertebrates with nearly 1850 species; they are unsegmented, pseudocoelomate, and bilaterally symmetric [4, 9, 10]. Nowadays, two classes are recognized: Pararotatoria (with the single order Seisonacea) and Eurotatoria, with two subclasses, Bdelloidea and Monogononta. The size of rotifers ranges from 50 to 2000 μm in length. Most of the species are free swimmers, but some species are fixed to some substratum [4]. Their morphology is of saccate type that is cylindrical with three easily recognizable regions: corona, trunk, and foot (Figure 1) [11].

1.2. Ecological relevance

Rotifers are cosmopolitans. They inhabit aquatic environments of the three types: marine, freshwater, and estuarine [9]. Most rotifers are freshwater, littoral with a few species truly planktonic. The few species that comprise the order Seisonacea and nearly 100 species of the subclass Monogononta are exclusively marine [4]. Species of the subclass Bdelloidea are found in freshwater ecosystems, lakes, temporary pools, interstitial water, soil, moss, and lichens [4, 12]. Rotifers are important in freshwater environments due to having one of the highest reproductive rate among metazoans, thus obtaining high population densities in short times, being dominant in many zooplanktonic communities. They act as links between the microbial community and the higher trophic levels. Rotifers colonize habitats quickly and convert

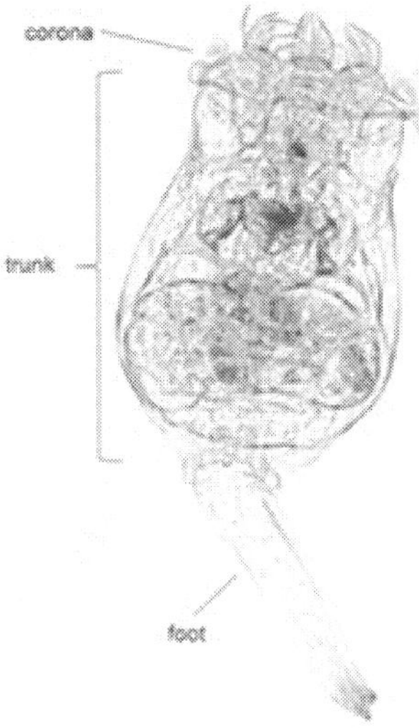

Figure 1. Diagram of the freshwater rotifer *Brachionuscalyciflorus* (Monogononta) [11].

primary production (algae and cyanobacteria) in a usable form for secondary consumers, making energy available for the next trophic levels. In interstitial water from swampy soils, they contribute to nutrient recycling [4]. Some bdelloids and Monogononta are abundant in wastewater treatment plants as part of activated sludge, in filtration systems, or residual lagoons feeding on the bacterial biomass [2, 13]. The population dynamics of rotifers is well characterized both in field as in the laboratory; for this reason, they are useful to investigate ecological and demographic principles. Besides, they are frequently used to assess aquatic toxicity from a population point of view. The intrinsic growth rate (r) is an end point commonly used and has proved to be highly sensitive [5, 14].

1.3. Culture

Rotifers can be obtained directly from a natural aquatic system or from laboratory cultures. Rotifers can be hatched from cysts, and today there are toxicity kits of two species: *B. calyciflorus* and *B. plicatilis* (Rotoxkits). The culture of rotifers has been developed for freshwater, estuarine, and marine organisms taken directly from the natural environment after a previous filtration, sterilization, and neutralization are performed. There is also synthetic hard like EPA medium [15] or synthetic marine water (Instant Ocean) that can be adjusted to the desired salinity. In freshwater species, usually the pH level is maintained in the physiological range (6.5–8) and temperature oscillates between 20 and 30°C. Marine species are maintained at salinities of 10 to 20 psu depending on a particular strain [16]. Rotifers are mainly filter feeders on microalgae, bacteria, or detritus; a few species are predatory. The supply of fresh and optimal food is the main problem for culturing rotifers. However, numerous rotifer species have been kept in the laboratory routinely [4]. Among the main marine microalgae used to feed rotifers are as follows: *Nannochloropsis* sp., *Chaetoceros* sp., *Dunaliella* sp., *Pyramimonas, Isochrysis* sp., and *Tetraselmis* sp.; for freshwater: *Nannochloris* sp., *Nannochloropsis* sp., and *Chlorella* sp. [16]. Culture media have to be renewed 2–3 times a week, and the supply of food must be frequent. Some species are maintained on diluted suspensions of commercial fish food, grain extracts or infusions, manure, and soil [4]. Among the most cultured species are found those of the genus *Brachionus*: *B. plicatilis, B. rotundiformis,* and *B. calyciflorus* [17, 18]. Some species of the genus *Lecane* can be easily cultured [19]. The economic importance of these rotifer species that are mass cultured are mainly based in their use as live feed for fish larvae and crustaceans in aquaculture [18].

1.4. Reproduction

The life cycles of rotifers are shorter than many other animals [14]. The Bdelloidea are reproduced by exclusive parthenogenesis and males are unknown. In the subclass Monogononta, the life cycle is haplodiploid with cyclic parthenogenesis as the dominant phase (amictic females), whereas the sexual reproduction (mixis) involves mictic diploid females that produce haploid eggs that if fertilized produce diploid embryos or cysts (in old literature they are called "resting eggs"), which go through diapause before hatching into amictic females that go back to the asexual phase of the cycle (Figure 2). Mixis is triggered by specific environmental stimuli like high population density or photoperiod [18, 20]. Parthenogenesis eliminates the high cost of producing males resulting in a rapid growth population [18], which allows for the high production of clonal individuals that can be used for aquaculture or toxicity tests.

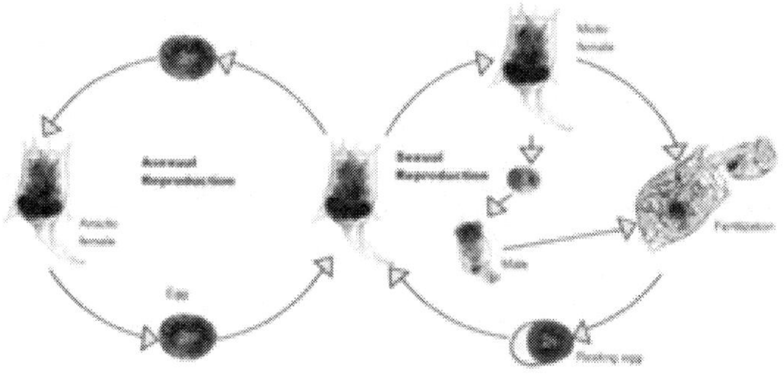

Figure 2. Reproductive cycle of *Brachionus calyciflorus*. Modified from Alvarado-Flores [21].

1.5. Cysts

Once the eggs are produced, they fall into the bottom and are deposited in the sediments [18]. The cysts are diploid and have a thick cover and can be viable for many years in dormancy. Cysts are very resistant to harsh environmental conditions like draught and freezing. Cysts can be dispersed through wind, water, or migratory animals. Under favorable conditions in a specific habitat, the cysts respond to a specific environmental clue (photoperiod, temperature, and salinity) and starts hatching producing an amictic diploid female [4]. Cysts can be stored for great periods of time without losing viability having rotifers in the desired period of time [22]. The use of rotifers hatched from cysts to develop toxicity tests was introduced by Snell and Persoone [23]. Nowadays, it is possible to obtain cysts from marine (*B. plicatilis*) and freshwater (*B. calyciflorus*) species in the market.

1.6. Rotifers as sentinels or bioindicators

Aquatic invertebrates are attractive model organisms in aquatic toxicology due to their short generation time compared with fishes besides their small size require small test volumes [14, 24]. Their small size, the fact that they are filter feeders, and the permeability of the integument made rotifers quite susceptible to chemical and physical changes [19, 25]. Due to the importance of rotifers in the aquatic trophic webs, they are useful as sentinel species to indicate toxicant exposure in affected ecosystems [4].

The knowledge of the (a) basic biology of rotifers, (b) sensitivity to contaminants at the physiological and demographic levels, (c) cosmopolitan distribution, (d) great ecological relevance, (e) high growth rate, (f) availability of neonates, (g) high ingestion rates, (h) ease of culture and handling, (i) transparency, (j) short life cycles, (k) ease to obtain clonal individuals, and (l) cyst production makes rotifers useful tools for assessment of aquatic ecosystems [4, 5, 8, 17, 18, 22, 24, 26].

2. Rotifers as model organism among invertebrates

Invertebrates are the most widely distributed organisms on the Earth and consist of a large and very diverse group, consisting in more than 30 phyla, several of which include more than 1000 different species [27]. The largest phylum within invertebrates is the Arthropoda, with more than one million species, in which insects and crustaceans are the two largest groups [28]. Aquatic crustaceans is a larger group in comparison to the aquatic insects, which a few of them have aquatic larvae [29]. Consequently, crustaceans are the most numerous and ecologically important group of invertebrates in marine and freshwater ecosystems, playing an important role in regulatory toxicity testing, whether in field or laboratory conditions [30, 31]. In addition, these methods are cost-effective because of the facility to obtain amictic eggs or in some cases resting eggs, which have been used to produce commercial kits for toxicity evaluation [32].

Current international protocols include several invertebrates' taxa for toxicity assessment (Table 1), in which cladocerans are the most used organisms and have more approved protocols for both acute (lethal) and (sub)chronic (sublethal) water toxicity assays than other taxa. The genus *Daphnia* is the most studied taxon within cladocerans. For the acute toxicity test, three replicates for at least five toxicant or effluent concentrations are required, every one of them with 100 mL, 48 h for exposure, without feeding [33]. Moreover, the chronic toxicity assay needs almost the same conditions but with feeding and daily medium renewal for 21 days [34]. In comparison, the *Brachionus* or other rotifer bioassays need only 1 mL per replicate [23], and the results can be obtained within 2 days for acute toxicity evaluation, and in 5 days for the chronic ones. Hence, rotifers tests involve similar conditions to the cladocerans, but when time is a matter of concern, the former is the best option to know the toxicity of chemicals or field samples.

Rotifers are even more versatile than daphnids. In April 2010, after the Macondo oil spill in the Gulf of Mexico, the Environmental Protection Agency of the United States (USEPA) in order to assess the impact of petroleum and oil dispersants on the aquatic biota recommended using *B. plicatilis,* a species complex that included *B manjavacas, B. plicatilis sensu stricto,* and *B. rotundiformis* among other that can exist [35]. This cost-effective test can be replicated hundreds of times in very short periods of time when necessary [36].

The Organization for the Economic Cooperation and Development (OECD) encourages developing test protocols that include crustaceans [37]. In this concern, the copepod *Eucyclops serrulatus* has been proposed as a model organism, with acute toxicity evaluations in 96 h [38] and chronic toxicity assessment in about 60 days [39]. These periods differ due to intrinsic characteristics of the species; copepods last longer to reach sexual maturity after hatching and require mating to obtain eggs for assays, while rotifer eggs are produced by parthenogenesis [40–42]. Nevertheless, with regard to the importance of copepods in aquatic ecosystems, there is not yet any approved protocol for toxicity evaluation as in rotifers [43, 44].

The cnidarian *Hydra attenuta* is another model organism included in standard methods [45]. This one is particularly interesting since lethal and sublethal end points are assessed in the same assay by quantifying the number of death animals and abnormalities found in the survivor hydras during 96 h for acute procedures, which are performed in 12-well microplates, whit a test volume of 4 mL [46]. Clubbed tentacles appearance is the first sign of abnormalities due to toxicants exposure. From this point, abnormalities follow shortening tentacles and the tulip stage (all tentacles have disappeared), and finally the disintegration of animal bodies occurs. Therefore, lethal effects should be registered from the tulip stage [47].

It is worth to compare with some rotifers in which it is also possible to determine sublethal effects in the same time period. Alvarado-Flores et al. [48] found abnormal body shapes in *B. calyciflorus* after vinclozolin exposure, demonstrating this pesticide also acts as an endocrine disruptor. As result, both models can be used for sublethal toxicity screening in short periods of time, but the difference lies on the apparent ease to determine such effects, while in *Hydra attenuata* are very obvious. In *B. calyciflorus,* more than one thousand organisms were analyzed to find statistically abnormalities in this rotifer species.

Bivalves in ecotoxicology procedures refer the use of species such as *Lampsilis siliquoidea, Corbicula fluminea, Leptodea fragilis, Ligumia subrostrata,* and *Megalonaias nervosa,* among others

[49, 50]. In the scientific literature, most of the bivalve specimens used in toxicological research and evaluations have been collected from environmental samples or commercial producers [51–54]. Glochida are dissected from adult bivalves and need a fish host to develop into juveniles [55–57]. Both immature stages, glochida and juveniles, are used for water quality assessment or chemicals released into aquatic systems [58]. In contrast, rotifer experimental organisms are produced in laboratory-controlled conditions, and animals are ready to use a few hours after hatching instead for several weeks to have juvenile unionids. Hence, rotifer bioassays are less time consuming than bivalves protocols.

In freshwater sediment, toxicity screening several protocols are currently carried out by international regulation agencies (Table 1). Nevertheless, rotifers are not included as test species, but it has been proved that the genus *Lecane* is suitable for those evaluations, representing a cost-effective and high reliable alternative to the standardized methods.

	Class	**Order**	**Family**	**Species**
			Freshwater	
Cladocera	Branchiopoda	Diplostraca	Daphniidae	*Daphnia magna* [62-65]
				D. pulex [66-68]
				D. similis [69]
				D. carinata [70-71]
				Ceriodaphnia dubia [62, 64, 66, [72]
Copepoda	Branchiopoda	Anostraca	Thamnocephalidae	*Thamnocephalus platyurus* [73-76]
	Maxillopoda	Calanoida	Cyclopidae	*Eurytemora affinis* [77-80]
				Eucyclops serrulatus [31, 81-83]
Rotifera	Monogononta	Ploima	Brachionidae	*Brachionus calyciflorus* [43, 84]
				B. rubens [23, 85-87]
				B. patulus [88-90]
			Lecanidae	*Lecane quadridentata* [91-93]
				L. hamata and L. luna [94,95]
				Euchlanis dilatata [96, 97]
			Asplanchnidae	*Asplanchna brightwelli* [98-101]
Cnidaria	Hydrozoa	Anthoathecata	Hydridae	*Hydra attenuata* [45]
				H. vulgaris [102-104]
Mollusca	Bivalvia	Unionoida	Unionidae	*Elliptio complanata* [58]
				Lampsilis siliquoidea [58]
		Veneroida	Corbiculidae	*Corbicula fluminea* [105-108]

	Class	Order	Family	Species
			Freshwater sediments	
Crustaceans	Malacostraca	Amphipoda	Hyalellidae	*Hyallela azteca* [109, 112]
Annelida	Clitellata	Lumbriculida	Lumbriculidae	*Lumbriculus variegatus* [113-115]
				Stylodrilus heringianus [116, 117]
Rotifera	Bdelloidea		Philodinidae	*Philodina roseola* [118]
Arthopoda	Insecta	Diptera	Chironomidae	*Chironomus dilutus* [109, 110, 119-121]
				C. riparius [109, 110, 119-121]
			Seawater	
Rotifera	Monogononta	Ploima	Brachionidae	*Brachionus plicatilis* [43]
Echinodermata	Echinoidea	Arbacioida	Arbaciidae	*Arbacia punctulata* [122]
Arthropoda	Malacostraca	Mysida	Mysidae	*Mysidopsis bahia* [123]
Copepoda	Maxillopoda	Calanoida	Acartiidae	*Acartia tonsa* [44, 124]
		Harpacticoida	Miraciidae	*Amphiascus tenuiremis* [44]
Molluscs	Bivalvia	Mytiloida	Mytilidae	*Mytilus edulis* [125, 126]
		Ostreoida	Osteridae	*Crassostrea gigas* [127]
			Seawater sediments	
Annelida	Polychaeta	Capitellida	Arenicolidae	*Arenicola marina* [128]
Artropoda	Malocostraca	Amphipoda	Corophiidae	*Leptocheirus plumulosus[110, 129, 130]*

Note: Numbers in square brackets indicate reference number

Table 1. Toxicity evaluations protocols using different invertebrates for freshwater and freshwater sediments, seawater, and seawater sediments.

Differently to the protocols mentioned above, all those for seawater and seawater sediments are found in standardized probes approved by agencies such as USEPA, ASTM, and the OECD (Table 1). As previously described, *B. plicatilis* is a marine rotifer used in seawater toxicology evaluation, an although it is not included in the USEPA guidelines, the agency recommended its used based on the ASTM method [43]. For sediments, the prime option is the polychaetes such as *Arenicola marina*, mainly collected from beach sand in different locations around the world. The protocol involves an exposure time of about 28 days using artificial sediment and field sediments samples [59–61]. Moreover, *B. plicatilis* has been exposed to interstitial water from littoral ecosystems; although it is sensitive to bioavailable toxicants in pore water, it does not respond in the same manner that those organisms that live in the sediment.

Afterward, the use of rotifers in all the above-mentioned matrices is possible due to some advantages, such as their relative small size, their simple organization, their short life cycles that permit multigenerational studies in very short time periods, their reproduction through parthenogenesis, and their genetic homogeneity that leads to almost identical offspring [22]. Despite of these characteristics, there is still a need to continue researching the effect of toxicants and to improve protocols for rotifers in (eco)toxicology to better understand their physiology and how external factors alter rotifers normal responses, since studies with rotifers are not as numerous as other taxa (Figure 3).

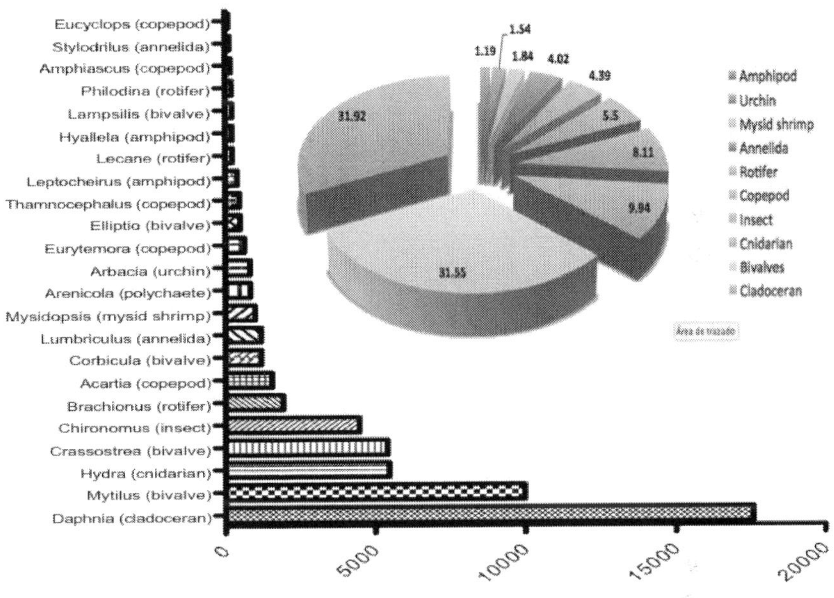

Figure 3. Number of publications for the groups described in the text. The databases consulted were Elsevier (www.sciencedirect.com), Springer (www.link.springer.com), and Wiley Online Library (http://onlinelibrary.wiley.com). Query terms were the genera name and the word "toxicity". Pie chart data represent percentages based on published material.

3. Rotifer sensitivity

The toxicity of a wide variety of chemicals including organic compounds, metals, and pharmaceuticals that can be found in water reservoirs as a result of human activity has been tested by performing toxicity tests using rotifers as bioindicators as these organisms play an important ecological role in aquatic environments [3, 131]. Rotifers of the genus *Brachionus* can be found between the most commonly used tests organisms [132]; moreover, several rotifers are included for comparison purposes.

A test organism with high sensitivity to detect adverse effects of xenobiotics is always desirable; nevertheless, differences in sensitivity among species can always be found. Generally, these are due to the biological characteristics of test organisms, the type of chemical, and its mode of action; thus, one species might be very sensitive to one toxic but not that sensitive to another compound [133]. This section focuses on the sensitivity of different rotifers, all of them from freshwater ecosystems, with exception of the marine species *B. plicatilis* and *Brachionus koreanus*, when testing the toxicity of pesticides, organic compounds, and metals attending their toxicological importance and the availability of literature.

Some examples of these variations on sensitivity are shown in Table 2. In relation to pesticides, malathion (organophosphorus) displays the lower toxicity and has a toxicity range from 33.72 to 45.5 mg/L, where the lower toxicity level is reported to *B. calyciflorus* and the higher to *B. plicatilis* [134, 135].

Lindane, an organochloride pesticide, and methyl parathion (organophosphorus) show moderate toxicity and have wider toxicity ranges in comparison to malathion. For the organochloride, toxicity range is given by *B. koreanus* and *B. plicatilis* (14 and 35.89 mg/L, respectively). In the case of methyl parathion, which shows higher acute toxicity, a range from 0.607 mg/L for *Euchlanis dilatata* to 29.19 mg/L for *B. calyciflorus* was found. Another organophosphorus of ecotoxicological relevance, chlorpyrifos, tends to affect significantly aquatic invertebrates due the LC50 values reported for *B. koreanus* (3.9 mg/L) and *B. calyciflorus* (11.85 mg/L) [96, 134, 136, 137].

The organochlorides endosulfan and pentachlorophenol (PCP) stand as highly toxic agrochemicals. For endosulfan, *B. koreanus* and *B. plicatilis* correspond to the reference organisms to establish the range because of their 24-h LC50 values found in literature [135, 136]. For PCP, the lowest and highest toxicity levels were noted for *B. calyciflorus* [138, 139].

On literature, little information is available about the toxicity of other organic compounds such as benzene, toluene, xylene, and hexane. Nevertheless, Ferrando and Andreu-Moliner [7] and Pérez-Legaspi et al. [95] have obtained some toxicological data (Table 2). For xylene, hexane, and toluene, *B. calyciflorus* and *B. plicatilis* are the reference organisms, where the first was most sensitive for the three chemicals [7, 95]. For benzene, rotifers from the genus *Lecane* resulted being more sensitive, but also a wide toxicity range is registered from 6.97 for *Lecane luna* to 3762 mg/L for *Lecane hamata* [95].

The lowest LC50 values, and thus, more toxic metals, were reported for zinc, copper, and mercury. Besides, a relatively not wide toxicity range can be observed for these three chemicals (Table 3). The most sensitive test organisms in these cases were *Lecane quadridentata*, *B. plicatilis*, and *B. calyciflorus*, respectively, while the more resistant were *Brachionus havanaensis* (Zn), *P. acuticornis* (Cu), and *L. hamata* (Hg) [93, 95, 135, 140, 143, 144].

Moreover, it is important to mention that there are physical factors that can alter the sensitivity of a test organism when performing a toxicity test. One example is given by Preston et al. [138],

Pollutant	Acute toxicity range in terms of 24h-LC50 (mg/L)
Pesticides	
Malathion	33.72[134] - 45.5[135]
Lindane	14[136] - 35.89[137]
Methyl parathion	*0.607[96] - 29.19[134]
Chlorpyrifos	3.9[136] – 11.85[137]
Endosulfan	4[136] - 6.6[135]
PCP	0.21[138] - 2.16[139]
Another organic compounds	
Xylene	252.7[7] - 495.9[7]
Hexane	68.3[7] - 154.3[7]
	Acute toxicity range in terms of 24 and 48h-LC[50](mg/L)
Benzene	6.97[95] – 3,762[95]
Toluene	113.3[7] - 552.6[7]

Note: Numbers in square brackets indicate reference number

*48h LC_{50} value

In the case of inorganic toxicants (Table 3), lead appears as the "less" toxic and with the wider toxicity range from 0.035 for *E. dilatata* to 56.2 mg/L for *Philodina acuticornis* [97, 140]. For cadmium, *E. dilatata* shows a great sensitivity with a 48-h LC50 of 0.014 mg/L, while *B. plicatilis* is more resistant (39 mg/L) [97, 141]. The most sensitive rotifer to chromium was *Lecane luna* (48-h LC50 of 3.26 mg/L) [95], and the highest 24-h LC50 of 17.4 mg/L was registered for *B. calyciflorus* [142].

Table 2. Toxicity range for some pesticides and organic compounds to rotifers used in bioassays.

Metal	Acute toxicity range in terms of 24 and 48h-LC[50] (mg/L)
Lead	0.035[97] - 56.2[140]
Cadmium	0.014[97] – 39[141]
Chromium	3.26[95] - 17.4[142]
Zinc	0.123[93] - 2.271[143]
Copper	0.01[135] - 1.9[140]
Mercury	0.06[144]- 1.37[95]

Note: Numbers in square brackets indicate reference number

Table 3. Toxicity range for different metals to rotifers used in bioassays.

where a decrease in LC50 values for PCP and mercury using *B. calyciflorus* as bioindicator was recorded as UV-B exposure time was increased.

Other physical condition that might modify the sensitivity of a test organism is desiccation. Robles-Vargas and Snell [145] found a remarkable resistance of *Philodina sp.* emerged from desiccation to PCP, chlorpyrifos, and mercury in comparison to continuously hydrated rotifers when reproductive tests were implemented [145].

According to the information reviewed and in terms of acute toxicity, rotifers from the family Brachionidae tend to show a great sensitivity to different chemicals, including organic and inorganic compounds. Invertebrates from the family Euchlanidae and Lecanidae appear to be sensitive mainly to inorganic toxicants. Furthermore, rotifers belonging to the family Philodinidae exhibit a notorious resistance to chemicals.

4. Rotifer database for screening toxicants

Every chemical product that can reach ecosystems and, as a consequence, the human being needs to be characterized to avoid noxious effects on all organisms exposed. At first instance, toxicological studies were mainly anthropocentrically focused, but that point of view would not have been enough to protect our own existence at all if affecting organisms in different ecosystems would have continued. It is known that rotifers play an important role as secondary producers and in some freshwater ecosystems represent the larger fraction of biomass of zooplankton [146]. Therefore, these organisms are nowadays considered as model organisms for toxicological screening.

Currently, one can find in the scientific literature reports of almost every class of chemical compound used or produced by man, investigating their toxicological properties. These papers are relatively abundant in several websites specialized in gathering such kind of information, which is available to everyone who has access to the published material. For scientist, collecting this material is a daily activity that guarantees their studies to follow the right direction by not working in exactly the same topic and manner, which would represent the duplicity of information but at higher cost because of multiple double efforts and resources investment. Nonetheless, it must be pointed out that differences in susceptibility to toxicants can be found among clones or strains, which is very interesting to assess by multiclonal analyses [22, 147]. However, when it is not possible to access multiple clones or strains, previous reports (published material) from databases become more relevant as a source of comparison to improve the comprehension of the observed results.

The USEPA, in an attempt to ease the search for environmental toxicological data, has created the ECOTOX database, hosted in their website [150]. It provides a user-friendly interface and allows discrimination among taxonomic and chemical groups, aquatic or terrestrial environment, tests results or conditions, and type (year) of publication. Although it does not give access to the cited references, it returns a list of publications to be consulted elsewhere. In the ECOTOX results table, the data shown are species name, exposure type, chemical, media type, end points, bioconcentration factor, effects, and statistics, among others (Figure 4). A complete list of all abbreviations used in the ECOTOX database can be directly downloaded for the same host website.

http://dx.doi.org/10.5772/61771

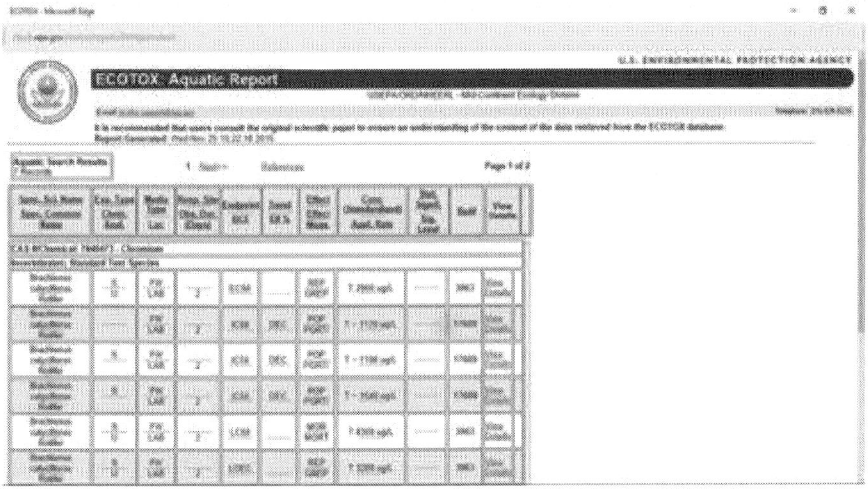

Figure 4. Print screen of ECOTOX database results. Query terms were *Brachionus* and chromium, selected from the options displayed in the database [150].

Nowadays, toxicity assays with rotifers include very diverse toxicants, and a relatively high number of test have been assessed the effect of inorganic (metals and metalloids) and very versatile organic compounds like pesticides, solvents, colorants, detergents, and emergent toxicants, in which health and care products and pharmaceuticals form part of it (Table 4). Some of these toxicants are effective at concentrations as low as some nanograms per liter, altering rotifers population dynamics by different mechanisms like endocrine disruption.

Species	Chemical compound or metal	Species	Chemical compound or metal
	1,1,1-trichloroethane [150]	Bc	Lauryl alcohol [150]
Bc	1,3,5-Trinitrobenzene [150]	Ab, Ai, Bc, Ed Lh, Ll, Lq, Pa	Lead [91, 94, 97]
	1-Dodecanesulfonic acid, Sodium salt [150]	Ap, Bc,Br, Kq	Lindane [150]
	1-Octanol [150]		Linuron [150]
Bp	2,2',4,4'-Tetrabromodiphenyl ether (BDE-47) [151]	Bc	Lithium [157]
Bc, Kc	2,3,4,6-Tetrachlorophenol [150]		m- and p-Xylene [150]
Ba, Ka,Kc	2,4,5-T [150]		Malathion [150]
Bc	2,4,6-trichlorophenol [150]	Br	Malathion [150]
	2,4,6-Trinitrotoluene [150]	Pa	Malathion [150]

Species	Chemical compound or metal	Species	Chemical compound or metal
	2,4-Dichloro aniline [150]	Bc	Malathion [152]
	2,4-Dichlorophenoxyacetic [150]	Bc	Malathion [153]
	2,4-dinitrochlorobenzene [150]	Bc, Li, Lq	Manganese [150], [166], [169]
	2,4-xylenol [150]	Bc	MARLON A390 [150]
	2,5-Dichloro aniline [150]	Bc, Bp, Ed, Lh, Ll, Lq, Pa	Mercury [91, 94, 97]
	2- [2- [2-(Dodecyloxy)ethoxy]ethoxy]ethanol, Hydrogen sulfate, Sodium salt [150]	Bc	Metallic silver [150]
	3,4-dichloro aniline [150]	Av	Methacrylamide [171]
	Acetaminophen [150]		Methacrylic acid [150]
Bc, Lh, Ll, Lq	Acetone [94, [150]]	Bc	Methanol [150]
Bc	Acetylsalicylic acid [152, 153]		Methoprene [150]
Av	Acrylamide [154]	As, Ba, Bc, Bp, Ed	Methyl parathion [97, 150]
	Acrylic acid [150]		Molinate [150]
Bc	Aldrin [155]		N,N,N-Trimethyl-1-octanamonium chloride [150]
	Alkyl* sodium benzene sulfonate [150]	Bc	N,N-Dimethyl-1-dodecanamine [150]
	Alkyl* trimethyl ammonium chloride [150]		N,N-Dimethyl-1-octanamine [150]
Ab, Bc, Li, Lq	Aluminum [93, 150, 156]		Naphthalenol [150]
Bc	Amitriptyline [152, 153, 157]		Naproxen [150]
Bc, Br	Ammonia [150]		n-Hexane [150]
Bc, Pa	Ammonium chloride [150]	Bc, Pa	Nickel [150]
Bc	Amphetamine sulfate [153]		Nicotine [152, 153]
Bc	Aroclor 1260 [150]		Nitric acid, Cadmium salt [150]
Bc, Pp, Pr,	Arsenic [150, 158>]		Nonyl phenol [150]
Bk	Atenolol [159]	Bc	Ofloxacin [161, 165]
Bc, Kq	Atrazine [150]		Orphenadrine [152, 153, 157]
Bc	Atropine [152, 153, 157]		Oxytetracycline [150]
Bc	Barium nitrate [150]	Bc, Bk	Oxytetracycline [159, 165]
Bc, Lh, Ll, Lq	Benzene [94, 150]	Bc	Paracetamol [152, 153]
Bc	Bezafibrate [160]		Para-chlorophenol [150]
Ab, Ai, Bc, Bh, Bp, Br, Bu, Ed, Lh, Ll, Lq, Pa	Cadmium [94, 143, 150, 160]	Bc	Para-dichlorobenzene [150]

Species	Chemical compound or metal	Species	Chemical compound or metal
Bc	Caffeine [150]		P-chloroaniline [150]
Pa	Calcium hypochlorite [150]	Ag, Bc, Br, Kq, Pa,	Pentachlorophenol PCP [150]
Bc, Bk	Carbamazepine [159, 161, 162]	Bc	Phenobarbital [152, 153, 157]
Ba, Hm, Mq, Pa	Carbaryl [150]	Bc, Br, Pa	Phenol [150]
Eb, Kq	Carbendazim [150]	Br	p-Nitrophenol [150]
	Carbofuran [150]	Bp	polybrominated diphenyl ethers [173]
Bc	Carbon tetrachloride [150]	Bc	Potassium chloride [150]
	Cetyl trimethyl ammonium chloride [150]	Bc, Pa	Potassium cyanide [150]
	Chloramphenicol [152, 153, 157]	Bc, Pa	Potassium dichromate [150]
Pa	Chlorine [150]		Prednisolone [174]
Bu	Chlornitrofen [150]		Prednisone [175]
	Chloroacetic acid [150]		Prococene [150]
Bc	Chloroform [150]		Propranolol [152, 153, 157, 176]
	Chloroquine [152, 153]	Bc	Quinidine [152]
Ab,Ba,Bb, Bc Bq, Bu, Eb, Ft, Hi, Hm, Kq, Mq, Pq, Tp	Chlorpyrifos [150,163]		Ranitidine [177]
Bc, Bp, Lh, Ll, Lq	Chromium [94, 150, 163]		Selenium [150]
Bc	Clarithromycin [165]	Pa	Silver nitrate [150]
Pa	Cobalt chloride [150]	Br	Simetryn [150]
Ab, Bc, Bp, Br, Lh, Ll, Lq, Li, Pa, Pp	Copper [94, 150, 158, 164, 166, 167]	Bp	SLS [178]
Ab, Bc, Bp	DDT, p,p' [150, 169]	Bc	Sodium bromide [150]
	Diazepam [152]	Bc	Sodium chloride [150]
	Diazinon [150]	Pr	Sodium chromate [150]
Bc	Diclofenac [161, 162]	Bc	Sodium dodecylbenzene sulfonate [150]
	Dicofol [150]	Pa	Sodium fluoride [150]
	Digoxin [152]	Bc	Sodium hypochlorite [150]
Pa	Dimethoate [150]	Bc	Sodium laurate [150]
Bc	Dinitro cresol [150]	Bc, Br	Sodium lauryl sulfate [150]

Species	Chemical compound or metal	Species	Chemical compound or metal
	Diphenylhydantoin [152]		Sodium oxalate [150]
	Endosulfan [150]		Sodium selenate [150]
	Endothall [150]		Sodium tetradecyl sulfate [150]
	Erythromycin [165]		Sulfamethoxazole [159, 161, 165]
	Estradiol [150]	Bc	Testosterone [168]
	Ethinylestradiol [94, 168]		Thallium(I) sulfate [150]
Lh, Ll, Lq	Ethyl acetate [94, 168]		Theophylline [152, 153]
	Ethyl alcohol [150]		ThiobencarB[150]
	Ethyl methacrylate [150]		Thioridazine [153]
Bc	Ethylene glycol [150]	Li	Tin [166]
	Fenitrothion [150]	Lh, Ll, Lq	Titanium [94]
	Fenofibrate [160]	Bc, Lh, Ll, Lq	Toluene [94, 150]
Bc, Pa	Ferrous sulfate [150]		Tributyl phosphate [150]
Ab	Iron [156]	Bc	Tributylstannane [150]
	Flutamide [150]		Tributyltin chloride [150]
Bc	Furosemide [170]	Bc, Kq. Pa	Trichlorfon [150]
	Gemfibrozil [160]	Bk	Trimethoprime [159]
Bc, Lq	Glyphosate [150]	Bc	Verapamil [152, 153]
Bc	Hexachlorophene [150]	Lh, Ll,Lq	Vinyl acetate [94]
	Hydroquinone [150]	Bc	Warfarin [152]
Ab, Li, Lq	Iron [96, 156, 166]		Xylene [150]
Bc	Isoniazid [152, 153]	Ab, Bc, Bh Li, Lq, Pa	Zinc [93, 143, 150, 156, 166]
	Isopropyl alcohol [150]		

Note: superscripts indicate reference number

Notes: Species are abbreviated with the first letters of their genus and species: Ad = *Adineta vaga*, Ab = *Asplanchna brightwellii*, Ag = *Asplanchna girodi*, Ai = *Asplanchna intermedia*, As = *Asplanchna sieboldi*, Ba = *Brachionus angularis*, Bb = *Brachionus bidentata*, Bc = *Brachionus calyciflorus*, Bh = *Brachionus havanaensis*, Bk = *Brachionus koreanus*, Bp = *Brachionus patulus*, Bp = *Brachionus plicatilis*, Bq = *Brachionus quadridentatus*, Br = *Brachionus Rubens*, Bu = *Brachionus urceolaris*, Eb = *Epiphanes brachionus*, Ed = *Euchlanis dilatata*, Ft = *Filinia terminalis*, Hi = *Hexarthra intermedia*, Hm = *Hexarthra mira*, Ka = *Keratella americana*, Kc = *Keratella cochlearis*, Kq = *Keratella quadrata*, Lh = *Lecane hamata*, Li = *Lecane inermis*, Ll = *Lecane luna*, Lq = *Lecane quadridentata*, Mq = *Monostyla quadridentata*, Pa = *Philodina acuticornis*, Pr = *Philodina roséola*, Pp = *Plationus patulus*, Pq = *Platyias quadricornis*, Tp = *Testudinella patina*.

Table 4. Relation of toxicants and rotifers species used in ecotoxicological studies

http://dx.doi.org/10.5772/61771

The available data about toxicity to rotifers show that organic compounds, as a complex group, is the most studied (32.4%) followed by the drugs group (24.58%), which could be due to the complexity of both groups, as they include a huge variety of toxicants. Inorganic compounds among other groups represent 17% out of the total number of tests included in this chapter, which can be split up into inorganic (nonmetallic compounds) and metals (Figure 5). The last one comprised those commonly named heavy metals [148]. Other groups might not be as numerous as the former ones, but this should not be misinterpreted as a lack of interest from the environmental toxicologists, but a challenge and a continuous research to find out how these chemicals are affecting the aquatic ecosystems and their inhabitants like rotifers species.

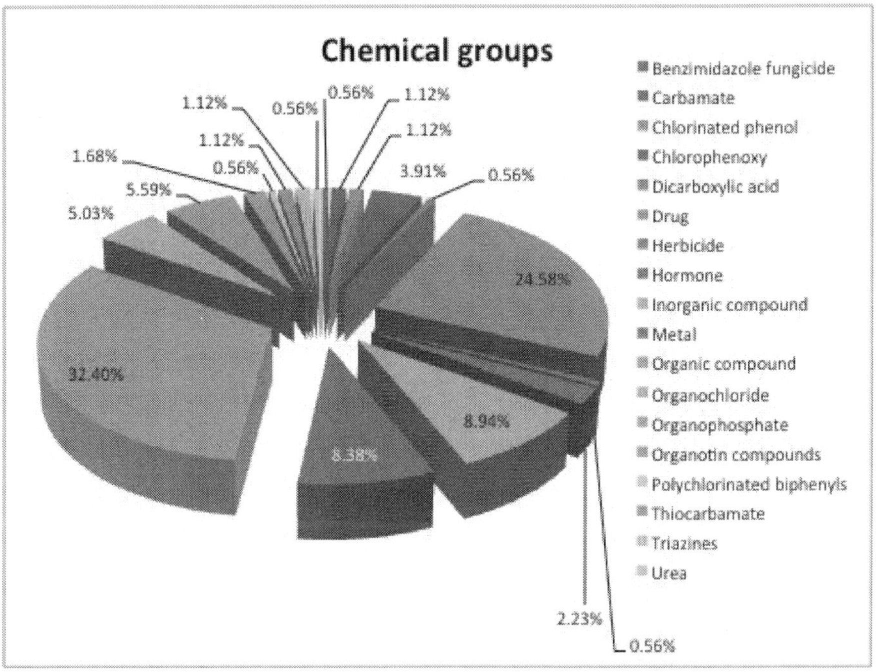

Figure 5. Comparison of toxicity tests between chemical groups conducted with different rotifers species.

Several rotifer species have been used to evaluate the effect and toxic mechanisms of the chemical groups aforementioned, where *B. calyciflorus* is the most abundant in the literature, with about 63% out of the total publications (Figure 6). However, there are other species that present advantages over brachionids, such as the littoral genus *Lecane* that might be used to assess the effect of sediment-associated toxicants, or the predator genus *Asplanchna*, used to evaluate toxicants biomagnification among rotifers and other aquatic invertebrates [149]. Therefore, rotifers represent a group of organisms that have not been completely studied and require more attention from toxicologists.

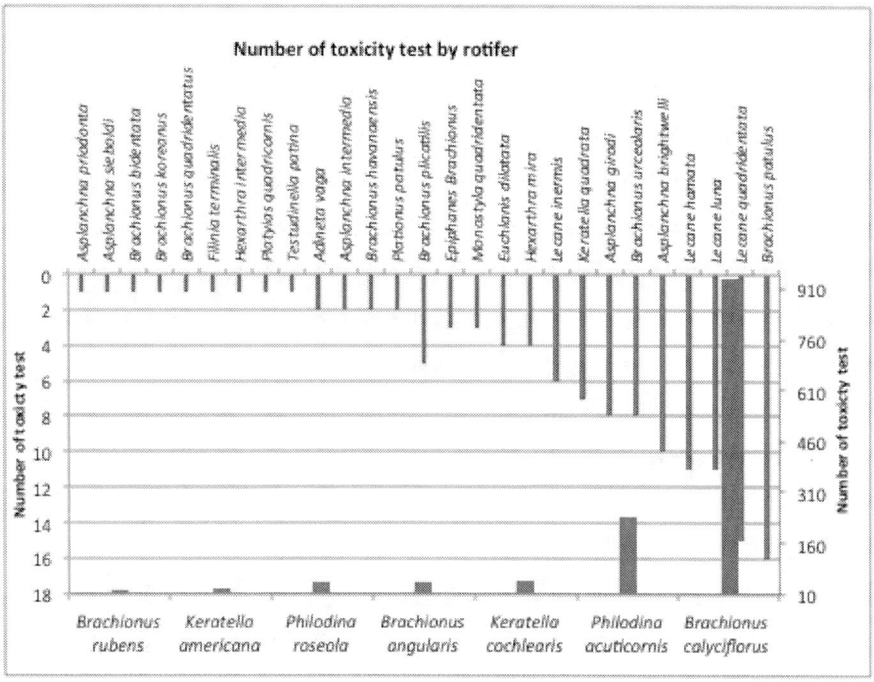

Red bars correspond to left y-axis and blue bars to right y-axis.

Figure 6. Number of ecotoxicological studies through by different rotifer species.

5. Rotifer tests already used worldwide and in certain regions

The ability to produce cysts has allowed the development of toxicity kits, called Rotoxkits, employed for acute/chronic marine and freshwater toxicity testing [179]. Cyst production is an outstanding characteristic that has enabled the development of several toxicity protocols using rotifers that have been used worldwide [180]. Rotifers are not directly represented in the legislation of several countries (as the cladocerans *Daphnia magna* and *Ceriodaphnia dubia* do). However, toxicity tests with rotifers have been published by official societies [43, 84, 181] Perhaps the most notorious participation of rotifer toxicity tests is when EPA asked BP plc (the former name of the company was British petroleum) to use the acute toxicity test with the euryhaline rotifer *B. plicatilis* to assess the toxicity generated after the 2010´s Gulf of Mexico oil spill [35]. The marine water rotifer toxicity test TK22 was used to analyze thousands of sites in the Gulf of Mexico; toxicity was analyzed for both oil and the oil dispersant used [36].

http://dx.doi.org/10.5772/61771

5.1. Europe

The Convention for the Protection of the Marine Environment of the North-East Atlantic (the "OSPAR Convention") was established in 1992. The convention started working in 1998. It has been ratified by Belgium, Denmark, Finland, France, Germany, Iceland, Ireland, Luxembourg, Netherlands, Norway, Portugal, Sweden, Switzerland, and the United Kingdom and approved by the European Community and Spain. One of their goals was the development of directives for the analysis of toxicity both in marine and freshwater environments. In their 2007 publication "Practical Guidance Document on Whole Effluent Assessment" (OSPAR, 2007), they applied the protocol of determination of the chronic toxicity to *B. calyciflorus* in 48 h [84] and validated its use.

In the Port of Aveiro, Portugal, standardized acute toxicity test using the marine rotifer, *B. plicatilis*, on sediment elutriates revealed higher toxicity levels in cases where the European Union Water Framework Directive suggested that sediment quality was acceptable [183]. Isidori et al. [184] employing *B. plicatilis* in 24-h toxicity tests found that all samples of municipal solid waste landfills in southern Italy expressed acute toxicity.

In Poland, the toxicity of the leachates from sewage sludge were investigated using different toxicological kits: Microtox (*Vibrio fischeri*), microbial assay for toxic risk assessment (ten bacteria and one yeast), Protoxkit F™ (*Tetrahymena thermophila*), Rotoxkit F™ (*B. calyciflorus*), and Daphtoxkit F™ (*D. magna*). Differences were observed in the sensitivity of the test organisms to the presence of sewage sludge in the soil. The highest sensitivity was a characteristic of *B. calyciflorus* [185].

5.2. Oceania

The Resource Management Act 1991 of New Zealand [186] requires local authorities and industry to apply for consent to discharge effluent to water or land. Therefore, the requirement for whole effluent toxicity testing is now being included in these Resource Consents. The Ministry for the Environment (MfE) has encouraged research which evaluates technology used to monitor environmental parameters (e.g., effluent toxicity). Such research was done using the Rotoxkit F™ and Rotoxkit M™ (with *B. calyciflorus* and *B. plicatilis*, respectively). The study concluded that "while very cost effective and with very good precision (repeatability), were not as sensitive as the others, and therefore are not recommended for whole effluent toxicity screening" [187].

5.3. Latin America and the Caribbean

Sarma et al. [188] showed that Mexico City urban wastewater affects instantaneous growth rate of *Brachionus patulus*. Acute 48-h lethal effect measurements generated with *L. quadridentata* on municipal, industrial, and agricultural sites around the city of Aguascalientes, Mexico, indicated that most samples tested were toxic [189]. The municipal drinking water wells of Aguascalientes have also been assessed using acute toxicity tests with *L. quadridentata* [190]. *L. quadridentata* has been used to assess the status of the major wastewater treatment plants in the State of Aguascalientes [191, 192]. An ecotoxicological study of the most important river in the State of Aguascalientes has also been performed [93]. *L. quadridentata* has also been used to assess toxicity in several rivers of the southern Huasteca area of Mexico with high concentrations of manganese (Mn) and the presence of DDT in the sediments and pore water. *L. quadridentata* was highly resistant to DDT and less susceptible to Mn than the cladocerans *D. magna* [165]. José de Paggi and Devercelli [193] examined the influence of watershed land use on microzooplankton around the city of Santa Fe in Argentina. Six rivers and a shallow lake located in rural and urban areas were sampled during 4 weeks. River microzooplankton abundance and rotifer species assemblages were found to be good indicators of land use. *Brachionus* spp. were associated with saline waters in rural areas and *Keratella* spp. (except *Keratella tropica*) with urban water bodies.

5.4. Asia

Many Asian countries have used rotifer toxicity tests for diverse monitoring and scientific purposes. Microcosm studies with rotifers have been used in India to evaluate tannery effluent [194]. *B. plicatilis* has been used to assess the toxicity of the various sewage sludge, one of the major ocean dumped materials in the Yellow Sea of Korea [195].

6. Perspectives for future studies regarding the importance of rotifers as models for toxicity screening of chemicals

Different rotifer species from all around the globe have been used to test the toxicity of a huge number of chemicals, both from freshwater (FW) and marine water (MW) ecosystems (see Rico-Martinez et al., 2013). Such species were initially collected from their natural habitats, a specific biogeographical zone, for their further acclimation to laboratory conditions and use as model organisms in toxicity evaluation protocols.

Nowadays, the rotifer species used as model organisms, due to their representativeness and wide natural distribution, include those of the genus *Brachionus* sp.: *B. calyciflorus* (FW), *B. plicatilis* (MW), *B. manjavacas* (MW), and *B. koreanus* (MW) (Lee et al.; Snell et al.), although some others like *B. rotundiformis* (MW) and *B. ibericus* (MW) could be used as model organisms (Pérez-Legaspi, 2015). Moreover, organisms within the genus *Lecane* sp. have been used for toxicological screening, demonstrating that *L. quadridentata* and other *Lecane* species are good indicators of water quality because of their high susceptibility to toxicants.

Despite all efforts to understand rotifers biology and their susceptibility to contaminants, there is still a need to conduct new studies with rotifers belonging to different habitats (biogeographical zones), climates, and niches. For such studies, researchers should take into consideration topics like clonal cultures obtaining, rotifers identification and classification through morphological and genetic (cytochrome oxidase rDNA, COI) characters, and the production of sexual eggs to preserve them in a resting eggs bank. Such eggs could be also a source for developing toxicity assessments kits, like those of Microbiotests Inc. [179], which mainly uses *Brachionus* sp. resting eggs in very efficient systems for their production.

Due to their immeasurable dispersion and diversity, annotating correctly the rotifers specimens' origin and recording their chemical backgrounds has become a very important issue because of the variable responses to toxicants observed within the same genus or even within the same species. For example, *Brachionus* sp. VER, isolated from the Gulf of Mexico, was the most tolerant to Macondo oil exposure (LC50 = 19.33%) in comparison to *B. rotundiformis* (LC50 = 11.02%) from Hawaii, *B. plicatilis* ss. (LC50 = 2.47%) from Tokyo, and *B. manjavacas* (LC50 = 5.43%) from Russia [36], which could be due to the presence of cryptic species within the taxon. Recently, researches and students from different countries participated in the workshop "Cryptic Speciation in *B. plicatilis*: A Workshop to Described Species within the Complex," and they estimated that there may be as many as 20 new species for this complex. In addition, the rotifers *B. calyciflorus* and *Lecane bulla* form a part of cryptic species [196, 197]. Thus, genetic and phylogeographic studies should be performed to assess how this species are distributed around the world.

In aquatic toxicology, currently there are methods that help elucidating the toxicity mechanisms for different sorts of chemicals. Therefore, they can be listed as follows:

a. Standardized protocols for evaluation of acute, chronic, and sublethal toxicity. These methods are carried out by exposing neonates or resting eggs, for 24 to 48 h or more depending exposure concentrations, periods, and end points to measure. Acute and chronic toxic ratios are still in use despite all new technologies, as they are finally a reference point for further analysis. Mortality or immobility are the common responses observed in acute toxicity tests, but in chronic assays, population parameters are followed during the exposure period, such as the intrinsic rate of population increase (r) obtained from the life table analysis. Another possibility is assessing the hatching percentage, which represent the stability and health of cysts produced during stressful conditions, including abiotic factors like desiccation or the presence of contaminants. Alvarado-Flores et al. [48] evaluated the effect of 1.2 mg/L of the fungicide vinclozolin on the rotifer *B. calyciflorus*. Their findings demonstrated that there was no significant difference between exposed and nonexposed organisms. In addition, the population parameter r of rotifers hatching from VZ-promoted cysts was 1.21 ± 0.063 (mean ± SD), and for rotifers hatching from control cysts was 0.90 ± 0.064.

It has been shown that multigenerational studies should be conducted as these could reproduce what happens in natural conditions when parthenogenetic females are exposed to toxicants and in their offspring is in certain way altered even before hatching [198], a phenomenon called the maternal effect, which could be for good while providing more energetic resources to deal with the stressful conditions in the medium, or negative through inheriting mutations that could bring deleterious effects in consecutive generations.

b. *In vivo* enzyme activity assessment (esterases and phospholipase A2): This method has the potential to assess the adverse effects of contaminants for rotifers. In the rotifer *E. dilatata* (FW), a native species from Mexico, the inhibition of esterases and phospholipase A2 was assessed fluorometrically after *in vivo* exposure (30 min) to sublethal concentrations of metals and pesticides in laboratory conditions. This study concluded that both enzymes are very sensitive to toxicants-induced alterations [96, 97].

c. Stress granules (SGs): Eukaryotes share diverse mechanisms of adaptation and responses to the stress. In this matter, it has been shown in different insects, trypanosomids, yeast, mammal cells, and rotifers, in which they can sequester some proteins and mRNA into granules that protect cellular mRNA. Thus, SGs appear to be useful as biomarkers in rotifers [199].

d. Bioconcentration factors (BCF): According to van der Oost et al. [200], biomarkers of exposure include the quantification of the toxicant or its metabolites. Therefore, BCF are very valuable tools to study exposure to a certain kind of compounds. Moreover, these assessments could help to trace toxicants in discharges to aquatic ecosystems, by monitoring both in laboratory and in natural conditions exposed animals [149, 201]

e. Elemental composition using X-ray analysis on rotifers cuticles. This is an easy method that qualitatively determines different elements of interest, principally inorganic metals [48].

f. Morphological analysis: These are changes induced by toxic exposure; although it could be controversial, it can be carried out by comparing morphological characters through image analysis. Because of rotifers phenotypic plasticity, the comparisons should be carefully performed to avoid misinterpretation of the results. However, is has been demonstrated, in *B. calyciflorus,* that morphological changes occurred after exposure to the fungicide vinclozolin and that abnormal and healthy animals are easily differentiated. However, the percentage of deformities is low, only 0.63% of 2868 organisms. Nevertheless, this is still significantly different to the control groups. Hence, morphometric analysis in rotifers could be a helpful tool to identify unrevealed targets of toxicants, and it might contribute to create a database for such effects and for several rotifers species to further comparisons among them, besides the likely identification process through image analysis [48].

g. Aging in rotifers: because there is a great diversity in aging rates among species, geographical populations, and mutants within species, Smith and Snell [202] designed an experiment to follow rotifers longevity through 84 generations (about 1 year). Their results show that optimal growing conditions (e.g., constant food supply) altered life span and can reduce aging, which could be evolutionary adjustable, with selection working primarily on the length of the reproductive life span. Rotifers are considered good models to investigate the effect not only of toxicant on their life span but also other factors such as caloric restriction and the effect of vitamins. Thus, this represents a new field to incorporate studies with rotifers [203, 204].

h. Hormones: Alvarado-Flores et al. [205] demonstrated the presence of some mammal-like proteins in the rotifer *B. calyciflorus* and concluded that it is necessary to generate more information about catecholaminergic and cholinergic systems in rotifers and the hormones related. Then it will be possible to assess their participation in mechanisms of detoxification and likely be used as toxicity models.

i. Genetic tools in rotifers and their applications

- RNA-seq: With mRNA-seq libraries for obligate parthenogenetic and cyclical parthenogenetic strains of the rotifer *B. calyciflorus*, it has been possible to identify genes specific to both modes of reproduction. Additionally, the studies performed by Hason et al. [206] allowed insights in the reproductive biology of obligated asexual bdelloid rotifers.

- Heat shock proteins (HSPs): The genes for these proteins synthesis are found from bacteria to higher eukaryotes and are related to functions like refolding denatured proteins due to stress that includes heat shock, reason for which they are called HSPs. Smith et al. [207] provided conclusive evidence that *hsp40, hsp60,* and *hsp70* are required for rotifer survival following heat stress, but that hsp90 seems not to be essential for survival, at least with their data.

- Metallothionein (Mt): These are low-molecular weight and cysteine-rich proteins present in eukaryotes. They provide potent metal binding and some other functions are being discovered. Their presence in rotifers has been demonstrated as a consequence of chromium exposure [208]

- P-glycoprotein 8P-gp (Pg-p): This protein could be considered as the first line of defense against some chemicals, including pharmaceuticals and endocrine disruptors. This protein has been found and characterized in the rotifer *B. koreanus*. Specimens of this rotifer were exposed to several pharmaceuticals that retarded growth and promoted the overexpression of Pg-p [209].

- Epigenetics: Germ cells can be specified early in embryogenesis by maternal determinants inherited in the cytoplasm of the oocyte or they can be selected later in the embryonic development from undifferentiated precursors by a localized inductive signal that is called epigenesis [210]. Epigenetic processes were found in the ovary of *B. plicatilis*, describing the participation of *vasa* and *nos* genes. As the first description of its kind, it opens the possibilities to explore and perform embryo development within the phylum.

- Cell-penetrating peptides (CPPs): Liu et al. (2013) demonstrated that some little peptides can penetrate cell membranes and delivered their cargoes; then their function (if cargoes have it) can be assessed as the functional HR9-delivered plasmid DNAs and RFP coding sequences that could be actively expressed in rotifers. This method provides a tool not only for genetic material but also for nanoparticles and proteins, which in the future could facilitate studying the effect of chemicals within rotifer cells.

j. Innate immunity in rotifers: In invertebrates, the nomenclature, annotation, and reports of cytokines could be controversial. Nonetheless, there is a continuous and increasing knowledge about cytokine-mediated immune regulation, although adaptive immune responses are likely absent in invertebrates, including rotifers. In this field, Jeong et al. [211] identified three genes of lipopolysaccharide-induced TNF-α factor (LITAF) in *B. koreanus*. The *in silico* analysis showed that these genes could be involved in innate immunity in primitive rotifers. In addition, exposure to lipopolysaccharide caused the overexpression of *LITAF1* and *LITAF2* but depleted glutathione concentration. Thereafter, *LITAF* genes have potential sensitivities to immune stimulator-triggered oxidative stress.

In conclusion, rotifers as models for ecotoxicological tests present several advantages, including a relative short life cycle that allows multigenerational studies and epigenetic research to unveil functions and processes in mictic and amictic rotifers, the simplicity of their body structure that ease the permeability of dyes for *in vivo* examination to quantify toxicant concentrations (e.g., Leadmium green®) or for systems descriptions like both the cholinergic and catecholaminergic systems, and their easy culture conditions and supply of resting eggs from different sources like some commercial brands. Nowadays, there are several protocols that describe the use of rotifers as indicators of water quality and safety as even thousands of probes can be performed in very short periods. Furthermore, every year, new technologies are becoming available to explore in deep detail the effect, the mechanisms, and the targets of toxicants. Hence, rotifer studies cannot be the exception, opening new possibilities to explore and describe more accurately the interaction of toxicants with the aquatic biota.

Author details

Roberto Rico-Martínez[1*], Mario Alberto Arzate-Cárdenas[1], Daniel Robles-Vargas[1], Ignacio Alejandro Pérez-Legaspi[2], Alvarado-Flores Jesús[3] and Gustavo Emilio Santos-Medrano[1]

*Address all correspondence to: rrico@correo.uaa.mx

1 Departamento de Química, Centro de Ciencias Básicas, Universidad Autónoma de Aguascalientes, Aguascalientes, AGS, Mexico

2 División de Estudios de Posgrado e Investigación, Instituto Tecnológico de Boca del Río, Boca del Río, Veracruz, Mexico

3 Catedrático CONACYT, Centro de Investigación Científica de Yucatán, Unidad de Ciencias del Agua, Cancún, Quintana Roo, México, Unidad de Ciencias del Agua, Cancún, Quintana Roo, Mexico

References

[1] Shane, B. Introduction to Ecotoxicology. In: Cockerham LG, Shane BS, editors. Basic Environmental Toxicology. Boca Raton, Florida: CRC Press; 1994. pp. 3–10.

[2] Rand, G.M. and Petrocelli, S.R. Fundamentals of Aquatic Toxicology. Hemisphere Publishing Corporation, New York. 1985. p. 666.

[3] Snell, T.W. and Janssen C.R. Rotifers in ecotoxicology: a review. Hydrobiologia. 1995. 313/314: 231–247. DOI: 10.1007/978-94-009-1583-1_32

[4] Wallace, R.L., Snell, T.W., Ricci, C. and Nogrady, T. Rotifera. 1: biology, ecology and systematics. In: Dumont HJF, editor. Guides to the Identification of the Microinvertebrates of the World. 2nd ed. The Netherlands: Kenobi Productions and Backhuys Publishers; 2006. p. 299.

[5] Pérez-Legaspi, I.A., Quintanar, J.L. and Rico-Martínez, R. Comparing toxicity endpoints on *Lecane quadridentata* (Rotifera: Monogononta) exposed to two anticholinesterases pesticides. Environment Toxicology. 2012; 27, 518–525. DOI: 10.1002/tox.20668

[6] Burbank, S. and Snell, T.W. Rapid toxicity assessment using esterase biomarkers in *Brachionus calyciflorus* (Rotifera). Environment Toxicology and Water Quality. 1994. 9: 171–178. DOI: 10.1002/tox.2530090303

[7] Ferrando, M.D. and Andreu-Moliner E. Acute toxicity of toluene, hexane, xylene and benzene to the rotifers *Brachionus calyciflorus* and *Brachionus plicatilis*. Bulletin of Environmental Contamination and Toxicology. 1992. 49: 266 –271. DOI: 10.1007/BF00191765

[8] Snell, T.W. and Janssen, C.R. Microscale toxicity testing with rotifers. In: Wells PG, Lee K, Blaise C, editors. Microscale Testing in Aquatic Toxicology, Advances, Techniques and Practice. Boca Raton, Florida: CRC Press. 1998. p 409–422.

[9] Pennak, R.W. Freshwater invertebrates of the United States. 3rd ed. New York: Wiley; 1989. 628 p.

[10] Segers, H. The nomenclature of the Rotifera: Annotated checklist of valid family-and genus-group names. Journal of Natural History. 2002. 36: 631–640. DOI: 10.1080/00222930010024573

[11] Pérez-Legaspi, I.A. Estudio de los efectos de dos pesticidas en el sistema colinérgico en especies selectas de rotíferos dulceacuícolas [thesis]. Aguascalientes: Universidad Autónoma de Aguascalientes; 2008.

[12] Clemént, P. and Wurdak, E. Rotifera. In: Harrison FW, Ruppert E, editors. Microscopy Anatomy of Invertebrates. Vol. 4: Aschelmintes. New York: Wiley-Liss; 1991. pp. 219–297.

[13] Klimek, B., Fialkowska, E., Kocerba-Soroka, W., Fida, J., Sobczyk, M. and Pajdak-Stós, A. The toxicity of selected trace metals to *Lecane inermis* rotifers isolated from activated sludge. Bulletin of Environmental Contamination of Toxicology. 2013. 91: 330–333. DOI: 10.1007/s00128-013-1062-z

[14] Snell, T.W. and Moffat, B.D. A 2-d life cycle test with the rotifer *Brachionus calyciflorus*. Environmental Toxicology and Chemistry. 1992. 11: 1249–1257. DOI: 10.1016/0147-6513(91)90070-6

[15] Peltier, W.H. and Weber, C.I., editors. Methods for measuring the acute toxicity of effluents to freshwater and marine organisms. EPA-600/4-85-013. Cincinnati, Ohio: US. Environmental Protection Agency; 1985. 231 p.

[16] Hoff, F.H. and Snell T.W. Plankton culture manual. 3rd ed. Dade City, Florida, USA: Florida Aqua Farms; 2008. 186 p.

[17] Yúfera, M. Studies on *Brachionus* (Rotifera): an example of interaction between fundamental and applied research. Hydrobiologia. 2001. 446/447: 383–392. DOI: 10.1023/A:1017583729646

[18] Kostopulou, V., Carmona, M.J. and Divanach P. The rotifer *Brachionus plicatilis*: an emerging bio-tool for numerous applications. Journal of Biological Research–Thessaloniki. 2012. 17: 97–112.

[19] Pérez-Legaspi, I.A. and Rico-Martínez R. Effect of temperature and food concentration in two species of littoral rotifers. Hydrobiologia. 1998. 387/388: 341-348. DOI: 10.1023/A:1017099906853

[20] Wallace, R.L. Rotifers: Exquisite metazoans. Integrative and Comparative Biology. 2002. 42: 660–667. DOI: 10.1093/icb/42.3.660

[21] Alvarado-Flores, J. Detección del efecto de disruptores endocrinos y metales en la reproducción de rotíferos dulceacuícolas. [thesis]. Aguascalientes: Universidad Autónoma de Aguascalientes; 2014.

[22] Dahms, H.U., Hagiwara, A. and Lee, J.S. Ecotoxicology, ecophysiology, and mechanistic studies wih rotifers. Aquatic Toxicology. 2011. 101 (1): 1–12. DOI: 10.1016/j.aquatox.2010.09.006

[23] Snell, T.W. and Persoone, G. 1989. Acute toxicity bioassays using rotifers. II. A freshwater test with *Brachionus rubens*. Aquatic Toxicology. 14 (1): 81–92. DOI: 10.1016/0166-445X(89)90056-8

[24] Janssen, C.R., Rodrigo, M.D.F. and Persoone, G. Ecotoxicological studies with the freshwater rotifer *Brachionus calyciflorus*. I. Conceptual framework and applications. Hydrobiologia. 1993. 255/256: 21–32. DOI: 10.1007/BF00025817

[25] Nogrady, T., Wallace, R.L. and Snell, T.W. Rotifera. In: Dumont, editor. Biology, Ecology and Systematics. Volume 4: Guides to the identification of the microinvertebrates of the continental waters of the world. The Netherlands: SPB Academic Publishing; 1993. p. 142.

[26] Marcial, H.S., Hagiwara, A. and Snell, T.W. Effect of some pesticides on reproduction of rotifer *Brachionus plicatilis* Müller. Hydrobiologia. 2005. 546: 569–575.

[27] Ruppert, E.E., Fox, R.S. and Barnes, R.D. Invertebrate zoology: a functional evolutionary approach, 7th ed. Thomson-Brooks/Cole, Belmont. 2004. pp. 617–625, 784–789

[28] Thorp, J.H. and Rogers, D.C. Field Guide to Freshwater Invertebrates of North America. Academic Press. Elsevier. San Diego, California. USA. 2010. pp. 109–119

[29] De Walt, R.W., Resh, V.H. and Hilsenhoff, W.L. Diversity and Classification or Insects and Collembola. In: Thorp JH. Ecology and Classification or North American Freshwater Invertebrates. Academic Press. Elsevier. San Diego. California. USA. 2010. pp. 587–593

[30] Hutchinson, T.H. Reproductive and developmental effects of endocrine disrupters in invertebrates: in vitro and in vivo approaches. Toxicology Letters. 2002. 131: 75–81. DOI:10.1016/S0378-4274(02)00046-2

[31] Baun, A., Hartmann, N.B., Grieger, K. and Kusk, K.O. Ecotoxicity of engineered nanoparticles to aquatic invertebrates: a brief review and recommendations for future toxicity testing. Ecotoxicology. 2008. 17(5):387–95. 10.1007/s10646-008-0208-y

[32] Cattaneo, A.G., Gornati, R., Chiriva-Internati, M. and Bernardini, G. Ecotoxicology of nanomaterials: the role of invertebrate testing. Invertebrate Survival Journal. 2009. 6: 78–97.

[33] U.S. EPA-821-R-02-012. Methods for measuring the acute toxicity of effluents and receiving waters to freshwater and marine organisms. 5th ed. 2002. http://water.epa.gov/scitech/methods/cwa/wet/upload/2007_07_10_methods_wet_disk2_atx.pdf

[34] U.S. EPA-821-R-02-013. Short-Term Methods for Estimating the Chronic Toxicity of Effluents and Receiving Water to Freshwater Organisms. 4th ed. 2002. http://water.epa.gov/scitech/methods/cwa/wet/upload/2007_07_10_methods_wet_disk3_ctf.pdf

[35] U.S. Environmental Protection Agency. Dispersant Monitoring and Assessment Final Directive for Subsurface Dispersant Application [Internet]. 2010. Available from: http://archive.epa.gov/bpspill/web/pdf/subsurface-dispersant-directive-final.pdf. [Accessed: 2015-09-03]

[36] Rico-Martínez, R., Snell, T.W. and Shearer, T.L. Synergistic toxicity of Macondo crude oil and dispersant Corexit 9500A(®) to the *Brachionus plicatilis* species complex (Rotifera). Environmental Pollution. 2013. 173: 5–10. DOI:10.1016/j.envpol.2012.09.024

[37] Gourmelon, A. and Ahtiainen, J. Developing Test Guidelines on invertebrate development and reproduction for the assessment of chemicals, including potential endocrine active substances—the OECD perspective. Ecotoxicology. 2007. 16: 161–167. DOI: 10.1007/s10646-006-0105-1

[38] Di Lorenzo, T., Di Marzio, W.D., Sáenz, M.E., Baratti, M., Dedonno, A.A., Iannucci, A., Cannicci, S., Messana, G. and Galassi, D.M. Sensitivity of hypogean and epigean freshwater copepods to agricultural pollutants. Environmental Science and Pollution Research International. Mar. 2014. 21 (6): 4643–455. DOI: 10.1007/s11356-013-2390-6

[39] Cifoni, M., Di Lorenzo, T., Faraloni, C. and Galassi D.M.P. A protocol for a development, reproduction and population growth test with freshwater copepods. Conference Paper. European Project AQUALIFE (LIFE12 BIO/IT/000231). 2015. International Conference on Groundwater in Karst. http://www.birmingham.ac.uk/generic/kgatb/index.aspx

[40] Gilbert, J.J. and Birky, C.W. Sensitivity and specificity of the *Asplanchna* response to dietary alpha-tocopherol. Journal of Nutrition. 1971. 101 (1): 113–126.

[41] Stelzer, C.P. Population regulation in sexual and asexual rotifers: an eco-evolutionary feedback to population size? Functional Ecology. 2012. 26: 180–188. DOI:10.1111/j.1365-2435.2011.01918.x

[42] Wallace., R.L. and Smith, H.A. Rotifera. In: eLS. John Wiley & Sons, Ltd: Chichester. 2013. DOI: 10.1002/9780470015902.a0001588.pub2

[43] ASTM E1440-91. Standard Guide for Acute Toxicity Test with the Rotifer Brachionus, ASTM International, West Conshohocken, PA. 2012. DOI: 10.1520/E1440-91R12. www.astm.org

[44] ASTM E2317-04. Standard Guide for Conducting Renewal Microplate-Based Life-Cycle Toxicity Tests with a Marine Meiobenthic Copepod, ASTM International, West Conshohocken, PA, 2012. DOI: 10.1520/E2317-04R12. www.astm.org

[45] ASTM STP921-EB. Aquatic Toxicology and Environmental Fate: Ninth Volume. 1986. DOI: 10.1520/STP921-EB. www.astm.org

[46] Quinn, B., Gagne, F. and Blaise, C. Validation of a multi-well plate toxicity test to assess feeding behavior of the cnidarian, *Hydra attenuata*. Fresenius Environmental Bulletin. 2007. 16: 1100–1107.

[47] Santos, M.A.P.F., Vicensotti, J. and Monteiro, R.T.R. Sensitivity of four test organisms (*Chironomus xanthus*, *Daphnia magna*, *Hydra attenuata* and *Pseudokirchneriella subcapitata*) to NaCl: an alternative reference toxicant. Journal of the Brazilian Society of Ecotoxicology. 2007. 2 (3): 229–236. DOI: 10.5132/jbse.2007.03.004

[48] Alvarado-Flores, J., Rico-Martínez, R., Adabache-Ortíz, A. and Silva-Briano, M. Morphological alterations in the freshwater rotifer *Brachionus calyciflorus* Pallas 1766 (Rotifera: Monogononta) caused by vinclozolin chronic exposure. Ecotoxicology. 2015. 24 (4): 915–925. DOI: 10.1007/s10646-015-1434-8

[49] Milam, C.D., Farris, J.L., Dwyer, F.J. and Hardesty, D.K. Acute toxicity of six freshwater mussel species (Glochidia) to six chemicals: implications for Daphnids and *Utterbackia imbecillis* as surrogates for protection of freshwater mussels (Unionidae). Archives of Environmental Contamination and Toxicology. 2005. 48 (2): 166–173. DOI: 10.1007/s00244-003-3125-3

[50] Besser, J.M., Ingersoll, C.G., Brumbaugh, W.G., Kemble, N.E., May, T.W., Wang, N., MacDonald, D.D. and Roberts, A.D. Toxicity of sediments from lead-zinc mining

areas to juvenile freshwater mussels (*Lampsilis siliquoidea*) compared to standard test organisms. Environmental Toxicology and Chemistry. 2015. 34 (3): 626–639. DOI: 10.1002/etc.2849

[51] Ringwood, A.H. Comparative sensitivity of gametes and early developmental stages of a sea urchin species (*Echinometra mathaei*) and a bivalve species (*Isognomon californicum*) during metal exposures. Archives of Environmental Contamination and Toxicology. 1992. 22 (3): 288–295. DOI: 10.1007/BF00212088

[52] Golding, C., Krassoi, R. and Baker, E. The development and application of a marine toxicity identification evaluation (TIE) protocol for use with an Australian bivalve. Australasian Journal of Ecotoxicology. 2006. 12: 37–44.

[53] Cheung, V.V., Depledge. M.H. and Jha, A.N. An evaluation of the relative sensitivity of two marine bivalve mollusc species using the Comet assay. Marine Environmental Research. 2006. 62 Suppl:S301-5. DOI: 10.1016/j.marenvres.2006.04.053

[54] Gerdol, M., De Moro, G., Manfrin, C., Milandri, A., Riccardi, E., Beran, A., Venier, P. and Pallavicini, A. RNA sequencing and *de novo* assembly of the digestive gland transcriptome in *Mytilus galloprovincialis* fed with toxinogenic and non-toxic strains of *Alexandrium minutum*. BMC Research Notes. 2014. 7: 722. DOI: 10.1186/1756-0500-7-722

[55] Akiyama, Y. and Iwakuma, T. Survival of glochidial larvae of the freshwater pearl mussel, *Margaritifera laevis* (Bivalvia: Unionoida), at different temperatures: a comparison between two populations with and without recruitment. Zoological Science. 2007. 24 (9): 890–893. DOI: 10.2108/zsj.24.890

[56] Dodd, B.J., Barnhart, M.C., Rogers-Lowery, C.L., Fobian, T.B. and Dimock, R.V. Jr. Persistence of host response against glochidia larvae in *Micropterus salmoides*. Fish and Shellfish Immunology. 2006. 21 (5): 473–484. DOI: 10.1016/j.fsi.2006.02.002

[57] Blakeslee, C.J., Galbraith, H.S., Robertson, L.S., and St John White, B. The effects of salinity exposure on multiple life stages of a common freshwater mussel, *Elliptio complanata*. Environmental Toxicology and Chemistry. 2013. 32 (12): 2849–2854. DOI: 10.1002/etc.2381

[58] ASTM E2455-06. Standard Guide for Conducting Laboratory Toxicity Tests with Freshwater Mussels. ASTM International, West Conshohocken, PA. 2013. DOI: 10.1520/E2455-06R13. www.astm.org

[59] Bat, L. and Raffaelli, D. Sediment toxicity testing: a bioassay approach using the amphipod *Corophium volutator* and the polychaete *Arenicola marina*. Journal of Experimental Marine Biology and Ecology. 1998. 226: 217–239. DOI: 10.1016/S0022-0981(97)00249-9

[60] Lewis, C., Pook, C. and Galloway, T. Reproductive toxicity of the water accommodated fraction (WAF) of crude oil in the polychaetes *Arenicola marina* (L.) and *Nereis virens* (Sars). Aquatic Toxicology. 2008. 90: 73–81. DOI: 10.1016/j.aquatox.2008.08.001.

[61] Besseling, E., Wegner, A., Foekema, E.M., van den Heuvel-Greve, M.J., Koelmans, A.A. Effects of microplastic on fitness and PCB bioaccumulation by the lugworm *Arenicola marina* (L.). Environmental Science and Technology. 2013. 47 (1): 593–600. DOI: 10.1021/es302763x.

[62] EPA 850.1010. Aquatic Invertebrate Acute Toxicity Test, Freshwater Daphnids. http://www.epa.gov/ocspp/pubs/frs/publications/OPPTS_Harmonized/850_Ecological_Effects_Test_Guidelines/Drafts/850-1010.pdf

[63] OECD. Test No. 202: *Daphnia* sp. Acute Immobilisation Test, OECD Guidelines for the Testing of Chemicals, Section 2, OECD Publishing, Paris. 2004. DOI: 10.1787/9789264069947-en

[64] ASTM E1295-01. Standard Guide for Conducting Three-Brood, Renewal Toxicity Tests with *Ceriodaphnia dubia*. ASTM International, West Conshohocken, PA. 2013. DOI: 10.1520/E1295-01R13 www.astm.org

[65] ISO 6341:2012. Water Quality. Determination of the Inhibition of the Mobility of *Daphnia magna* Straus (Cladocera, Crustacea). Acute Toxicity Test. International Organization for Standardization. Geneva, Switzerland. 2012. http://www.iso.org/iso/iso_catalogue/catalogue_tc/catalogue_detail.htm?csnumber=54614

[66] EPA 850.1300. Short-Term Methods for Estimating the Chronic Toxicity of Effluents and Receiving Waters to Freshwater Organisms. 4th ed. 2002. http://water.epa.gov/scitech/methods/cwa/wet/upload/2007_07_10_methods_wet_disk3_ctf.pdf

[67] OECD. Test No. 211: *Daphnia magna* Reproduction Test, OECD Guidelines for the Testing of Chemicals, Section 2. OECD Publishing, Paris. 2012. DOI: 10.1787/9789264185203-en

[68] ASTM E1193-97. Standard Guide for Conducting *Daphnia magna* Life-Cycle Toxicity Tests. ASTM International, West Conshohocken, PA. 2012. DOI: 10.1520/E1193-97R12. www.astm.org

[69] CETESB. Métodos de avaliação da toxicidade de poluentes a organismos aquáticos. São Paulo. 1992.

[70] NSW EPA. Cladoceran toxicity testing. Laboratory Procedures Manual. 1999. 5: 17.

[71] NSW EPA. Chronic reproduction impairment test. Laboratory Procedures Manual. 2003. 6: 16

[72] ISO 20665. Water quality. Determination of chronic toxicity to *Ceriodaphnia dubia*. International Organization for Standardization. Geneva, Switzerland. 2008. http://www.iso.org/iso/catalogue_detail.htm?csnumber=39673

[73] Sepic, E., Bricelj, M. and Leskovsek, H. Toxicity of fluoranthene and its biodegradation metabolites to aquatic organisms. Chemosphere. 2003. 52 (7): 1125–1133. DOI: 10.1016/S0045-6535(03)00321-7

[74] Heinlaan, M., Ivask, A., Blinova, I., Dubourguier, H.C. and Kahru, A. Toxicity of nanosized and bulk ZnO, CuO and TiO_2 to bacteria *Vibrio fischeri* and crustaceans *Daphnia magna* and *Thamnocephalus platyurus*. Chemosphere. 2008. 71 (7): 1308–1316. DOI: 10.1016/j.chemosphere.2007.11.047.

[75] Brausch, J.M. and Smith, P.N. Mechanisms of resistance and cross-resistance to agrochemicals in the fairy shrimp *Thamnocephalus platyurus* (Crustacea: Anostraca). Aquatic Toxicology. 2009. 92 (3): 140–145. DOI: 10.1016/j.aquatox.2009.02.002

[76] Blinova, I., Niskanen, J., Kajankari, P., Kanarbik, L., Käkinen, A., Tenhu, H., Penttinen, O.P. and Kahru, A. Toxicity of two types of silver nanoparticles to aquatic crustaceans *Daphnia magna* and *Thamnocephalus platyurus*. Environmental Science and Pollution Research International. 2013. 20 (5): 3456–3463. DOI: 10.1007/s11356-012-1290-5.

[77] Ger KA, Teh SJ, Goldman CR. Microcystin-LR toxicity on dominant copepods *Eurytemora affinis* and *Pseudodiaptomus forbesi* of the upper San Francisco Estuary. Science of the Total Environment. 2009. 407 (17): 4852–4857. DOI: 10.1016/j.scitotenv.2009.05.043

[78] Cailleaud, K., Budzinski, H., Lardy, S., Augagneur, S., Barka, S., Souissi, S. and Forget-Leray, J. Uptake and elimination, and effect of estrogen-like contaminants in estuarine copepods: an experimental study. Environmental Science and Pollution Research International. 2011. 18 (2): 226–236. DOI: 10.1007/s11356-010-0355-6

[79] Michalec, F.G., Holzner, M., Menu, D., Hwang, J.S. and Souissi, S. Behavioral responses of the estuarine calanoid copepod *Eurytemora affinis* to sub-lethal concentrations of waterborne pollutants. Aquatic Toxicology. 2013. 15: 138–139:129–138. DOI: 10.1016/j.aquatox.2013.05.007

[80] Leseur, T., Boulangé-Lecomte, C., Restoux, G., Deloffre, J., Xuereb, B., Le Menach, K., Budzinski, H., Petrucciani, N., Marie, S., Petit, F. and Forget-Leray, J. Toxicity of sediment-bound pollutants in the Seine estuary, France, using a *Eurytemora affinis* larval bioassay. Ecotoxicology Environmental Safety. 2015. 113:169–175. DOI: 10.1016/j.ecoenv.2014.11.033

[81] Sánchez-Bayo, F. Comparative acute toxicity of organic pollutants and reference values for crustaceans. I. Branchiopoda, Copepoda and Ostracoda. Environmental Pollution. 2006. 139 (3): 385–420. DOI: 10.1016/j.envpol.2005.06.016

[82] Nandini, S. and Sarma, S.S.S. Effect of algal and animal diets on life history of the freshwater copepod *Eucyclops serrulatus* (Fischer, 1851). Aquatic Ecology. 2007. 41 (1): 75–84. DOI: 10.1007/s10452-006-9051-5

[83] Di Lorenzo, T., Di Marzio, W.D., Spigoli, D., Baratti, M., Messana, G., Cannicci, S. and Galassi, D.M.P. Metabolic rates of a hypogean and an epigean species of cope-

pod in an alluvial aquifer. Freshwater Biology. 2015. 60: 426–435. DOI: 10.1111/fwb. 12509

[84] ISO 20666. Water quality. Determination of the chronic toxicity to *Brachionus calyciflorus* in 48 h. International Organization for Standardization. Geneva, Switzerland. 2008. http://www.iso.org/iso/catalogue_detail.htm?csnumber=39674

[85] Arauzo, M. and Valladolid, M. Short-term harmful effects of unionised ammonia on natural populations of *Moina micrura* and *Brachionus rubens* in a deep waste treatment pond. Water Research. 2003. 37 (11): 2547–2554. DOI: 10.1016/S0043-1354(03)00023-X

[86] Azuara-García, R., Sarma, S.S.S. and Nandini S. The combined effects of zinc and alga on the life table demography of *Anuraeopsis fissa* and *Brachionus rubens* (Rotifera). Journal of Environmental Science and Health. Part A, Toxic/Hazardous Substances and Environmental Engineering. 2006. 41 (4): 559–572. DOI: 10.1080/10934520600564394

[87] Geng, H. and Xie, P. Experimental studies on the effects of toxic *Microcystis aeruginosa* PCC7820 on the survival and reproduction of two freshwater rotifers *Brachionus calyciflorus* and *Brachionus rubens*. Ecotoxicology. 2008. 17 (8): 709–715. DOI: 10.1007/s10646-008-0219-8

[88] Sarma, S.S.S., Ramírez-Pérez, T. and Nandini, S. Comparison of the sensitivity of *Brachionus calyciflorus* and *Brachionus patulus* (Rotifera) to selected heavy metals under low and high food (*Chlorella vulgaris*) levels. Bulletin of Environmental Contamination and Toxicology. 2000. 64 (5): 735–739. DOI: 10.1007/s001280000065

[89] Sarma, S.S.S., Nandini, S. and Pérez, T.R. Combined effects of mercury and algal food density on the population dynamics of *Brachionus patulus* (Rotifera). Bulletin of Environmental Contamination and Toxicology. 2001. 67 (6): 841–847. DOI: 10.1007/s001280199

[90] Peredo-Alvarez, V.M., Sarma. S.S. and Nandini, S. Combined effect of concentrations of algal food (*Chlorella vulgaris*) and salt (sodium chloride) on the population growth of *Brachionus calyciflorus* and *Brachionus patulus* (Rotifera). Revista de Biología Tropical. 2003. 51 (2): 399–407.

[91] Hernández-Flores, S. and Rico-Martínez, R. Study of the effects of Pb and Hg toxicity using a chronic toxicity reproductive 5-day test with the freshwater rotifer *Lecane quadridentata*. Environmental Toxicology. 2006. 21 (5): 533–540. DOI: 10.1002/tox. 20218

[92] Domínguez-Cortinas, G., Mejía-Saavedra, J., Santos-Medrano, G.E. and Rico-Martínez, R. Analysis of the toxicity of glyphosate and Faena® using the freshwater invertebrates *Daphnia magna* and *Lecane quadridentata*. Toxicological and Environmental Chemistry. 2008. 90 (2): 377–384. DOI: 10.1080/02772240701529038

[93] Torres-Guzmán, F., Avelar-González, F.J. and Rico-Martínez, R. Implementing *Lecane quadridentata* acute toxicity tests to assess the toxic effects of selected metals (Al, Fe

and Zn). Ecotoxicology and Environmental Safety. 2010. 73 (3): 287–295. DOI: 10.1016/j.ecoenv.2009.10.006

[94] Pérez-Legaspi, I.A. and Rico-Martínez, R. Acute toxicity tests on three species of the genus Lecane (Rotifera: Monogononta). Hydrobiologia. 2001. 446/447 (1): 375–381. DOI: 10.1023/A:1017531712808

[95] Pérez-Legaspi, I.A., Rico-Martínez, R. and Pineda-Rosas, A. Toxicity testing using esterase inhibition as a biomarker in three species of the genus *Lecane* (Rotifera). Environmental Toxicology and Chemistry. 2002. 21 (4): 776–782. DOI: 10.1002/etc.5620210413

[96] Arias-Almeida, J.C., Rico-Martínez, R. Inhibition of two enzyme systems in *Euchlanis dilatata* (Rotifera: Monogononta) as biomarker of effect of metals and pesticides. Biomarkers. 2011. 16 (1): 12–9. DOI: 10.3109/1354750X.2010.520336

[97] Arias-Almeida, J.C. and Rico-Martínez, R. Toxicity of cadmium, lead, mercury and methyl parathion on *Euchlanis dilatata* Ehrenberg 1832 (Rotifera: Monogononta). Bulletin of Environmental Contamination and Toxicology. 2011. 87 (2): 138–1342. DOI: 10.1007/s00128-011-0308-x

[98] Gómez, A., Cecchine, G. and Snell, T.W. Effect of pentachlorophenol on predator-prey interaction of two rotifers. Aquatic Toxicology. 1997. 37 (4): 271–282. DOI: 10.1016/S0166-445X(96)00833-8

[99] Gama-Flores, J.L., Ferrara-Guerrero, M.J., Sarma, S.S. and Nandini, S. Prey (Brachionus calyciflorus and *Brachionus havanaensis*) exposed to heavy metals (Cu and Cd) for different durations and concentrations affect predator's (*Asplanchna brightwellii*) population growth. Journal of Environmental Science and Health. Part A, Toxic/Hazardous Substances and Environmental Engineering. 2007. 42 (10): 1483–1488. DOI: 10.1080/10934520701480912

[100] Guo, R., Snell, T.W. and Yang, J. Studies of the effect of environmental factors on the rotifer predator–prey system in freshwater. Hydrobiologia. 2010. 655 (1): 49–60. DOI: 10.1007/s10750-010-0403-8

[101] Chen, J., Wang, Z., Li, G. and Guo, R. The swimming speed alteration of two freshwater rotifers *Brachionus calyciflorus* and *Asplanchna brightwelli* under dimethoate stress. Chemosphere. 2014. 95: 256–260. DOI: 10.1016/j.chemosphere.2013.08.086

[102] Karntanut, W., Pascoe, D. A comparison of methods for measuring acute toxicity to *Hydra vulgaris*. Chemosphere. 2000. 41 (10): 1543–1548. DOI: 10.1016/S0045-6535(00)00068-0

[103] Pascoe, D., Carroll, K., Karntanut, W. and Watts, M.M. Toxicity of 17alpha-ethinylestradiol and bisphenol A to the freshwater Cnidarian *Hydra vulgaris*. Archives of Environmental Contamination and Toxicology. 2002. 43 (1): 56–63. DOI: 10.1007/s00244-001-0016-3

[104] Pascoe, D., Karntanut, W. and Müller, C.T. Do pharmaceuticals affect freshwater invertebrates? A study with the cnidarian *Hydra vulgaris*. Chemosphere. 2003. 51 (6): 521–528. DOI: 10.1016/S0045-6535(02)00860-3

[105] Cataldo, D., Colombo, J.C., Boltovskoy, D., Bilos, C. and Landoni, P. Environmental toxicity assessment in the Paraná river delta (Argentina): simultaneous evaluation of selected pollutants and mortality rates of *Corbicula fluminea* (Bivalvia) early juveniles. Environmental Pollution. 2001. 112 (3): 379–389. DOI: 10.1016/S0269-7491(00)00145-7

[106] Achard, M., Baudrimont, M., Boudou, A., Bourdineaud, J.P. Induction of a multixenobiotic resistance protein (MXR) in the Asiatic clam *Corbicula fluminea* after heavy metals exposure. Aquatic Toxicology. 2004. 67 (4): 347–357. DOI: 10.1016/j.aquatox.2004.01.014

[107] Cherry, D.S., Scheller, J.L., Cooper, N.L. and Bidwell, J.R. Potential effects of Asian clam (*Corbicula fluminea*) die-offs on native freshwater mussels (Unionidae) I: water-column ammonia levels and ammonia toxicity. Journal of the North American Benthological Society. 2005. 24 (2): 369–380. DOI: 10.1899/04-073.1

[108] Liao, C.M., Jou, L.J., Lin, C.M., Chiang, K.C., Yeh, C.H. and Chou, B.Y. Predicting acute copper toxicity to valve closure behavior in the freshwater clam *Corbicula fluminea* supports the biotic ligand model. Environmental Toxicology. 2007. 22 (3): 295–307. DOI: 10.1002/tox.20263

[109] ASTM E1706-05. Standard Test Method for Measuring the Toxicity of Sediment-Associated Contaminants with Freshwater Invertebrates. ASTM International, West Conshohocken, PA. 2010. DOI: 10.1520/E1706-05R10. www.astm.org

[110] EPA OPPTS 850.1735. Whole Sediment Acute Toxicity Invertebrates, Freshwater. 1996. http://www.epa.gov/ocspp/pubs/frs/publications/OPPTS_Harmonized/850_Ecological_Effects_Test_Guidelines/Drafts/850-1735.pdf

[111] EPA 600/R99.064 Methods for Measuring the Toxicity and Bioaccumulation of Sediment-associated Contaminants with Freshwater Invertebrates. 2000. http://water.epa.gov/polwaste/sediments/cs/freshfact.cfm

[112] EPS 1/RM/33. Environment Canada. Biological Test Method: Test for Survival and Growth in Sediment Using the Freshwater Amphipod *Hyalella azteca*. 2012. 184 p. http://ec.gc.ca/Publications/4BA9E8C1-30EA-48AE-B593-F8F763EED8F0/RM33---accessible.PDF

[113] ASTM E1688-10. Standard Guide for Determination of the Bioaccumulation of Sediment-Associated Contaminants by Benthic Invertebrates, ASTM International, West Conshohocken, PA, 2010. DOI: 10.1520/E1688-10. www.astm.org

[114] EPA 823-f-00-002. Methods for Measuring the Toxicity and Bioaccumulation of Sediment-associated Contaminants with Freshwater Invertebrates—Second Edition Fact Sheet. 2000. 2 p. http://www.epa.gov/ostwater/cs/freshfact.html

[115] OECD *Test No. 225:* Sediment–Water *Lumbriculus Toxicity* Test Using Spiked Sediment, OECD Guidelines for the Testing of Chemicals. Section 2. OECD Publishing. Paris. 2007. DOI: 10.1787/9789264067356-en

[116] White, D.S. and Keilty, T.J. Burrowing avoidance assays of contaminated Detroit River sediments, using the freshwater Oligochaete *Stylodrilus heringianus* (Lumbriculidae). Archives of Environmental Contamination and Toxicology. 1998. 17 (5): 673–681. DOI: 10.1007/BF01055837

[117] Keilty, T.J., White, D.S. and Landrum, P.F. Sublethal responses to endrin in sediment by *Stylodrilus heringianus* (Lumbriculidae) as measured by a 137cesium marker layer technique. Aquatic Toxicology. 1988. 13 (3): 251–269. DOI: 10.1016/0166-445X(88)90056-2

[118] Moreira, R.A., da Silva Mansano, A. and Rocha, O. 2015. The toxicity of carbofuran to the freshwater rotifer, *Philodina roseola*. Ecotoxicology. 24 (3): 604–615. DOI: 10.1007/s10646-014-1408-2

[119] EPA OPPTS 850.1790. Chironomid sediment toxicity test. Ecological Effects Test Guidelines. 1996.http://www.epa.gov/ocspp/pubs/frs/publications/OPPTS_Harmonized/850_Ecological_Effects_Test_Guidelines/Drafts/850-1790.pdf

[120] OECD. Test No. 235: *Chironomus* sp., Acute Immobilisation Test. OECD Guidelines for the Testing of Chemicals, Section 2. OECD Publishing, Paris. 2011. DOI: 10.1787/9789264122383-en

[121] EPS 1/RM/32. Environment Canada. Biological Test Method: Test for Survival and Growth in Sediment using the Larvae of Freshwater Midges *Chironomus tentans* or *Chironomus riparius*. 1997. 150 p. http://publications.gc.ca/site/eng/460451/publication.html

[122] EPA 821-R-02-014 Short-term Methods for Estimating the Chronic Toxicity of Effluents and Receiving Waters to Marine and Estuarine Organisms. 3rd ed. 486 p. http://water.epa.gov/scitech/methods/cwa/wet/upload/2007_07_10_methods_wet_disk1_ctm.pdf

[123] EPA/600/R-09/121 *Americamysis bahia* Stochastic Matrix Population Model for Laboratory Populations: Technical Documentation. 2009. http://nepis.epa.gov/Exe/ZyPDF.cgi/P1005OBH.PDF?Dockey=P1005OBH.PDF

[124] ISO 16778:2015. Water quality. Calanoid copepod early-life stage test with *Acartia tonsa*. International Organization for Standardization. Geneva, Switzerland. 2015. http://www.iso.org/iso/catalogue_detail.htm?csnumber=57698

[125] ASTM E2122-02. Standard Guide for Conducting In-situ Field Bioassays with Caged Bivalves. ASTM International, West Conshohocken, PA, 2013. DOI: 10.1520/E2122-02R13. www.astm.org

[126] ASTM E724-98. Standard Guide for Conducting Static Acute Toxicity Tests Starting with Embryos of Four Species of Saltwater Bivalve Molluscs. ASTM International, West Conshohocken, PA. 2012. DOI: 10.1520/E0724-98R12. www.astm.org

[127] EPA 850.1055. Bivalve acute toxicity test (embryo larval). Ecological Effects Test. 1996. 7p. http://www.epa.gov/ocspp/pubs/frs/publications/OPPTS_Harmonized/850_Ecological_Effects_Test_Guidelines/Drafts/850-1055.pdf

[128] Thain, J. and Bifield, S. Biological effects of contaminants: Sediment bioassay using the polychaete *Arenicola marina*. ICES Techniques in Marine Environmental Sciences, 2001. 29: 16 pp.

[129] ASTM E1367-03. Standard Test Method for Measuring the Toxicity of Sediment-Associated Contaminants with Estuarine and Marine Invertebrates. ASTM International, West Conshohocken, PA. 2014. DOI: 10.1520/E1367-03R14. www.astm.org

[130] EPA 600-R-01-020. Methods for Assessing the Chronic Toxicity of Marine and Estuarine Sediment-associated Contaminants with the Amphipod *Leptocheirus plumulosus*. 1st ed. 2001. 121 pp. http://water.epa.gov/polwaste/sediments/cs/upload/guidance-manual.pdf

[131] Zadorozhnaya, O., Kirsanov, D., Buzhinsky, I., Tsarev, F., Abramova, N., Bratov, A., Javier-Muñoz, F., Ribó, J., Bori, J., Carmen-Riva, M. and Legin, A. Water pollution monitoring by an artificial sensory system performing in terms of *Vibrio fischeri* bacteria. Sensors and Actuators B. 2015. 207: 1069–1075. DOI: 10.1016/j.snb.2014.08.056

[132] Janssen, C. R., Ferrando-Rodrigo, M. D. and Persoone, G. Ecotoxicological studies with the freshwater rotifer *Brachionus calyciflorus*. I. Conceptual framework and applications. Hydrobiologia. 1993. 255/256: 21–32. DOI: 10.1007/978-94-011-1606-0_4

[133] Vaal, M., van der Wal, J. T., Hermens, J. and Hoekstra, J. Pattern analysis of the variation in the sensitivity of aquatic species to toxicants. Pergamon. 1997. 35 (6): 1291–1309. DOI: 10.1016/S0045-6535(97)00166-5

[134] Fernández-Casalderrey, A., Ferrando, M. D. and Andreu-Moliner, E. Acute toxicity of several pesticides to rotifer (*Brachionus calyciflorus*). Bulletin of Environmental Contamination and Toxicology. 1992a. 48: 14–17. DOI: 10.1007/BF00197477

[135] PAN Pesticides Database. Chemical Toxicity Studies on Aquatic Organisms. Aquatic toxicity. Available from: http://www.pesticideinfo.org/ [Accessed: 2015-08-10]

[136] Kim, B. M., Lee, J. W., Seo, J. S., Shin, K. H., Rhee, J. S. and Lee J. S. Modulated expression and enzymatic activity of the monogonont rotifer *Brachionus koreanus* Cu/Zn- and Mn-superoxide dismutase (SOD) in response to environmental biocides. Chemosphere. 2015. 120: 470–478. DOI: 10.1016/j.chemosphere.2014.08.042

[137] Ferrando, M. D. and Andreu-Moliner, E. Acute lethal toxicity of some pesticides to *Brachionus calyciflorus* and *Brachionus plicatilis*. Bulletin of Environmental Contamination and Toxicology. 1991. 47: 479–484. DOI: 10.1007/BF01702214

[138] Preston, B. L., Snell, T. W. and Kneisel, R. UV-B exposure increases acute toxicity of pentachlorophenol and mercury to the rotifer *Brachionus calyciflorus*. Environmental Pollution. 1999. 106: 23–31. DOI: 10.1016/S0269-7491(99)00065-2

[139] Ferrando, M. D., Andreu-Moliner, E. and Fernández-Casalderrey, A. Relative sensitivity of *Daphnia magna* and *Brachionus calyciflorus* to five pesticides. Journal of Environmental Science and Health. Part B. 1992. 27 (5): 511–522. DOI: 10.1080/03601239209372798

[140] Buikema, A. L., Cairns, J. J. and Sullivan, G. W. Evaluation of *Philodina acuticornis* (Rotifera) as a bioassay organism for heavy metals. Water Resources Bulletin. 1974. 10 (4): 648–661. DOI: 10.1111/j.1752-1688.1974.tb05624.x

[141] Snell, T. W., Moffat, B. D., Janssen C. and Persoone, G. Acute toxicity tests using rotifers. III. Effects of temperature, strain and exposure time on the sensitivity of *Brachionus plicatilis*. Environmental Toxicology and Water Quality. 1991. 6 (1): 63–75. DOI: 10.1002/tox.2530060106

[142] Sarma, S. S. S., Martínez-Jerónimo, F., Ramírez-Pérez, T. and Nandini, S. Effect of cadmium and chromium toxicity on the demography and population growth of *Brachionus calyciflorus* and *Brachionus patulus* (Rotifera). Journal of Environmental Science and Health. Part A. 2006. 41: 543–558. DOI: 10.1080/10934520600564311

[143] Juárez-Franco, M. F., Sarma, S. S. S. and Nandini, S. Effect of cadmium and zinc on the population growth of *Brachionus havanaensis* (Rotifera: Brachionidae). Journal of Environmental Science and Health. Part A. 2007. 42: 1489–1493. DOI: 10.1080/10934520701480938

[144] Snell, T. W., Moffat, B. D., Janssen, C. and Persoone, G. Acute toxicity tests using rotifers: IV. Effects of cyst age, temperature and salinity on the sensitivity of *Brachionus calyciflorus*. Ecotoxicology and Environmental Safety. 1991. 21(3): 308–317. DOI: 10.1016/0147-6513(91)90070-6

[145] Robles-Vargas, D. and Snell, T. W. Effects of desiccation on the toxicant sensitivity of rotifers. Hydrobiologia. 2010. 652: 185–193. DOI: 10.1007/s10750-010-0330-8

[146] Makarewicz, J. C. and Likens, G. E. 1979. Structure and function of the zooplankton community of Mirror Lake, New Hampshire. Ecological Monographs. 49: 109–127.

[147] Sun, D., Ma, R., Liu, W. and Niu, C. Sexual reproduction and short-term fitness advantage in the rotifer *Brachionus calyciflorus*: implications for the coexistence of sympatric clones. Chinese Journal of Oceanology and Limnology. 2013. 31 (5): 987–993. DOI: 10.1007/s00343-013-2271-3

[148] Chapman, P. Heavy metal-music, not science. Environmental Science and Technology. 2007. 41 (12): 6C.

[149] Rubio-Franchini, I. and Rico-Martínez, R. Evidence of lead biomagnification in invertebrate predators from laboratory and field experiments. Environmental Pollution. 2011. 159 (7): 1831–1835. DOI: 10.1016/j.envpol.2011.03.021.

[150] ECOTOX Database Query. Advanced Database Query. 2015. Available from: http://cfpub.epa.gov/ecotox/advanced_query.htm [Accesed: 2015-09-03].

[151] Wang, H., Tang, X., Sha, J., Chen, H., Sun, T, and Wang, Y. The reproductive toxicity on the rotifer *Brachionus plicatilis* induced by BDE-47 and studies on the effective mechanism based on antioxidant defense system changes. Chemosphere. 2015. 135: 129–37. DOI: 10.1016/j.chemosphere.2015.03.090

[152] Calleja, M., C., Personne, G. and Geladi, P. Comparative acute toxicity of the first 50 multicentre evaluation of in vitro cytotoxicity chemicals to aquatic non-vertebrates. Archives of Environmental Contamination and Toxicology.1994. 26:69–78. DOI: 10.1007/BF00212796

[153] Calleja, M., C., Personne, G and Geladi, P. Human acute toxicity prediction of the first 50 MEIC chemicals by a battery of ecotoxicological tests and physicochemical properties. Food and Chemical Toxicology. 1994. 32 (2): 173–187. DOI: 10.1016/0278-6915(94)90180-5

[154] Orstan, A. Toxicity of Acrylamide Derivatives to Embryos of the Rotifer *Adineta vaga*. Bulletin of Environmental Contamination and Toxicology. 1992. 48 (6): 901–906. DOI: 10.1007/BF00201152

[155] Huang, L., Xi, Y., Zha, W. and Wen, X. Effect of aldrin on life history characteristics of rotifer *Brachionus calyciflorus* Pallas. Bulletin of Environmental Contamination and Toxicology. 2007. 79: 524–528. DOI: org/10.1051/limn/2013038

[156] Santos-Medrano, G.E., and Rico-Martinez, R. Lethal effects of five metals on the freshwater rotifers *Asplanchna brightwellii* and *Brachionus calyciflorus*. Hidrobiológica. 2013. 23 (1): 82–86.

[157] Calleja, M.C., Persoone, G. and Geladi, P. The predictive potential of a battery of ecotoxicological tests for human acute toxicity as evaluated with the first 50 MEIC chemicals. ATLA Alternative to Laboratory Animals. 1993. 21 (3): 330–349.

[158] Rios-Arana, V., J., Gardea-Torresdey, J., L., Webb, R. and Walsh, E., J. Heat shock protein 60 (HSP60) response of *Plationus patulus* (Rotifera:Monogononta) to combined exposures of arsenic and heavy metals.. Hydrobiologia. 2005. 546: 577–586. DOI: 10.1007/1-4020-4408-9_58

[159] Rhee, J., Jeong, C., B., Kin, B., M. and Lee. J., L. P-glycoprotein (P-gp) in the monogonont rotifer, *Brachionus koreanus*: molecular characterization and expression in response to pharmaceuticals. Aquatic Toxicology 2012. 114–115: 104–118. DOI: 10.1016/j.aquatox.2012.02.009

[160] Isidori, M., Nardelli, A., Pascarella, L., Rubionp. M. and Parrella. A. Toxic and genotoxic impact of fibrates and their photoproducts on non-target organisms. Environmental International. 2007. 33: 635–641. DOI: 10.1016/j.envint.2007.01.006

[161] Ferrari, B., Paxéus, N., Lo Giudice, N., Pollio, A. and Garric, J. Ecotoxicological impact of pharmaceuticals found in treated wastewaters: study of carbamazepine, clofibric acid, and diclofenac. Ecotoxicology and Environmental Safety. 2003. 55: 359–370. DOI: 10.1016/S0147-6513(02)00082-9

[162] Ferrari, B., Mons, R., Vollat, B., Fraysse, B., Paxéus, N., Lo Giudice, R., Pollio, A., Garric, J. Environmental risk assessment of six human pharmaceuticals: are the current environmental risk assessment procedures sufficient for the protection of the aquatic environment? Environmental Toxicology and Chemistry. 2004. 23 (5): 1344–1354. DOI: 10.1897/03-246

[163] Hurlbert, S.H., M.S. Mulla, and H.R. Willson. Effects of an organophosphorus insecticide on the phytoplankton, zooplankton, and insect populations of fresh-water ponds. Ecological Monographs. 1972. 42(3): 269–299. DOI: 10.2307/1942211

[164] Filenko, O. F. and. Samoylova, T. A. Effect of heavy metals on samples and model population of rotifers *Brachionus plicatilis* Muller at different salinity. Moscow University Biological Sciences Bulletin. 2008. 63 (2): 80–83. DOI: 10.3103/S0096392508020077

[165] Isidori, M., Lavorgna, M., Nardelli, A., Pascarella, L. and Parrella, A. Toxic and genotoxic evaluation of six antibiotics on non-target organisms. Science of the Total Environment. 2005. 346: 87–98. DOI: 10.1016/j.scitotenv.2004.11.017

[166] Klimek, B., Fiałkowska, E., Kocerba-Soroka, W., Fyda, J., Sobczyk, M. and Pajdak-Stós, A. The toxicity of selected trace metals to *Lecane inermis* rotifers isolated from activated sludge. Bulletin of Environmental Contamination and Toxicology. 2013. 91:330–333. DOI: 10.1007/s00128-013-1062-z

[167] Enesco, H.E., Wolanskyj, A. and Sawada, M. Effect of copper on lifespan and lipid peroxidation in rotifers. Age. 1989. 12 (1): 19–23. DOI: 10.1007/BF02432514

[168] Radix, P., Severin, G., Schramm, K-W., and Kettrup, A. Reproduction disturbances of *Brachionus calyciflorus* (rotifer) for the screening of environmental endocrine disrupters. Chemosphere. 2002. 47: 1097–1101. DOI: 10.1016/S0045-6535(01)00335-6

[169] Mejía-Saavedra, J., Sánchez-Armass, S., Santos-Medrano, G., E., Gonzáles-Amaro, R., Razo-Soto, I., Rico-Martínez, R. and Díaz-Barriga, F. Effect of coexposure to ddt and manganese on freshwater invertebrates: pore water from contaminated rivers and laboratory studies. Environmental Toxicology and Chemistry. 2005. 24 (8): 2037–2044. DOI: 0.1897/04-438R

[170] Isidori, M., Nardelli, A., Parrella, A., Pascarella, L. and Previtera, L. A multispecies study to assess the toxic and genotoxic effect of pharmaceuticals: furosemide and its photoproduct. Chemosphere. 2006. 63: 785–793.

[171] Orstan, A. Toxicity of acrylamide derivatives to embryos of the rotifer *Adineta vaga*. Bulletin of Environmental Contamination and Toxicology. 1992. 48 (6): 901–906.

[172] Isidori, M., Lavorgna, M., Nardelli, A., Parella, A., Previtera, L. and Rubino, M. Ecotoxicity of naproxen and its phototransformation products. Science of the Total Environment. 2005. 348: 93–101. DOI: 10.1016/j.scitotenv.2004.12.068

[173] Sha, J., Wang, Y., Chen, H., Wang, M., Wang, H., Li, X., Qi L., and Tang, X. Using population demographic parameters to assess impacts of two polybrominated diphenyl ethers (BDE-47, BDE-209) on the rotifer *Brachionus plicatilis*. Ecotoxicology and Environmental Safety. 2015. 119: 106–115. DOI: 10.1016/j.ecoenv.2015.05.009

[174] DellaGreca, M., Fiorentino, A., Isidori, M., Lavorgna, M., Previtera, L., Rubino, M. and Temussi, F. Toxicity of prednisolone, dexamethasone and their photochemical derivatives on aquatic organisms. Chemosphere. 2004. 54: 629–637. DOI:10.1016/j.chemosphere.2003.09.008

[175] DellaGreca, M., Fiorentino, A., Lesce, M.m R., Isidori, M., Nardelli, A., Previtera, L. and Temussi, F. Identification of phototransformation products of prednisone by sunlight: toxicity of the drug and its derivatives on aquatic organisms. Environmental Toxicology and Chemistry. 2003. 22 (3): 534–539. DOI: 10.1002/etc.5620220310

[176] Liu, Q.T., Williams, T.D., Cumming, R.I., Holm, G., Hetheridge, M.J. and Murray-Smith, R. Comparative aquatic toxicity of propranolol and its photodegraded mixtures; algae and rotifer screening. Environmental Toxicology and Chemistry. 2009. 28: 2622–2631. DOI: 10.1897/09-071.1

[177] Isidori, M., Parrella, A., Pistillo, P. and Temussi, F. Effects of ranitidine and its photoderivatives in the aquatic environment. Environment International. 2009. 35:821–825. DOI: 10.1016/j.envint.2008.12.002

[178] Persoone, G., Van de Vel, A., Van Steertegem, M. and De Nayer, B. Predictive value of laboratory tests with aquatic invertebrates: influences of experimental conditions. Aquatic Toxicology. 1989. 14: 149–166. DOI: 10.1016/0166-445X(89)90025-8

[179] MicroBioTests Inc. [Internet]. 2015. Available: http://www.microbiotests.be [Accessed 2015-09-03]

[180] Rico-Martínez, R., Pérez-Legaspi, I.A., Arias-Almeida, J.C and Santos-Medrano, G.E. 2013. Rotifers in ecotoxicology. In: Encyclopedia of Aquatic Ecotoxicology (Férard J. F., y Blaise C., Eds.). Springer, Berlin, Germany, 973–996. DOI: 10.1007/SpringerReference_380004

[181] APHA. Standard Methods for the Examination of Water and Wastewater. 20th ed. American Public Health Association, Washington, DC. 1998.

[182] OSPAR Commission. Practical Guidance Document on Whole Effluent Assessment. Publication number 316/2007. OSPAR Commission. 2007.

[183] Gonçalves, S.F., Calado, R., Gomes, N.C.M.., Soares, A.M.V.M. and Loureiro, S. An ecotoxicological analysis of the sediment quality in a European Atlantic harbor emphasizes the current limitations of the Water Framework Directive. Marine Pollution Bulletin. 2013. 72 (1): 197–204. DOI: 10.1016/j.marpolbul.2013.04.003

[184] Isidori, M., Lavorgna, M., Nardelli, A., Parrella, A. Toxicity identification evaluation of leachates from municipal solid waste landfills: a multispecies approach. Chemosphere. 2003. 52 (1): 85–94. DOI: 10.1016/S0045-6535(03)00298-4

[185] Malara A. and Oleszczuk, P. Application of a battery of biotests for the determination of leachate toxicity to bacteria and invertebrates from sewage sludge-amended soil. Environmental Science and Pollution Research International. 2013. 20 (5): 3435–3446. DOI: 10.1007/s11356-012-1268-3

[186] Resource Management Act. Public Act 1991 No 69. New Zealand Legislation. Parliamentary Counsel Office. [Internet]. 1991. Available from: http://www.legislation.govt.nz/act/public/1991/0069/latest/whole.html#DLM230265 [Accessed: 2015-08-20].

[187] Ruck, J.G., Martin, M. and Mabon, M. 2000. Evaluation of Toxkits as methods for monitoring water quality in New Zealand. In: New Microbiotests for Routine Toxicity Screening and Biomonitoring. Ed. Persoone, G., Janssen, C. and De Coen, W. 103–119 p. DOI: 10.1007/978-1-4615-4289-6_9

[188] Sarma, S.S.S., Trujillo-Hernández, H.E. and Nandini, S. Population growth of herbivorous rotifers and their predator (*Asplanchna*) on urban wastewaters. Aquatic Ecology. 2003. 37: 243–250. DOI: 10.1023/A:1025896703470

[189] Santos-Medrano, G.E., Ramírez-López, E.M., Hernández-Flores S. Azuara-Medina, P.M. and Rico-Martínez, R. Determination of toxicity levels in the San Pedro River watershed, Aguascalientes, Mexico. Journal of Environmental Science and Health. Part A. Toxic/Hazardous Substances and Environmental Engineering. 2007. 42: 1403–1410. DOI: 10.1080/10934520701480557

[190] Rico-Martínez R., Velázquez-Rojas, C.A., Pérez-Legaspi, I.A. and Santos-Medrano, G.E. 2001. The use of aquatic invertebrate toxicity tests and invertebrate enzyme biomarkers to assess toxicity in the states of Aguascalientes and Jalisco, Mexico. In: Butterworth, F.M., Gunatilake, A. and Gonsebatt-Bonaparte M.E. Eds. Biomonitors and Biomarkers as Indicators of Environmental Change, Volume 2. Plenum Press, Nueva York. pp. 427–438.

[191] Torres-Guzmán, F., Avelar-González, F.J., y R. Rico-Martínez. An assessment of chemical and physical parameters, several contaminants including metals, and toxicity in the seven major wastewater treatment plants in the state of Aguascalientes, Mexico. Journal of Environmental Science and Health. Part A. Toxic/Hazardous Sub-

stances and Environmental Engineering. 2010. 45 (1): 2–13. DOI: 10.1080/10934520903388517

[192] Robles-Vargas D., S. M. Montoya-Castillo, F. J. Avelar-González, J. Jauregui-Rincón, F. J. Rodríguez-Valadez, & R. Rico-Martínez. Assessment of the quality and toxicity of the discharges of a wastewater treatment plant and alternatives to improve its operation. Journal of Environmental Health and Science. Part A. Toxic/Hazardous Substances and Environmental Engineering. 2012. 47 (4): 589–597. DOI: 10.1080/10934529.2012.650578

[193] José de Paggi, S., and Devercelli, M. Land use and basin characteristics determine the composition and abundance of the microzooplankton. Water, Air and Soil Pollution. 2011. 218 (1/4): 93–108. DOI: 10.1007/s11270-010-0626-3

[194] Koteswari, Y.N. and Ramanibai, R. Evaluation of toxicity of tannery effluent on plankton community structure: a multispecies microcosm study II. Turkish Journal of Biology. 2004. 28: 55–63.

[195] Park, G., S., Lee, S.H., Park, S.Y., Yoon, S.J., Lee, S.M., Chung, C.S., Hong, G.H. and Kim S. H. Ecotoxicological evaluation of sewage sludge using bioluminescent marine bacteria and rotifer. Ocean Science Journal. 2005. 40 (2): 91–100. DOI: 10.1007/BF03028589

[196] Gilbert, J.J. and Walsh, E.J. *Brachionus calyciflorus* is a species complex: mating behavior and genetic differentiation among four geographically isolated strains. Hydrobiologia. 2005. 546: 257–265. DOI: 10.1007/s10750-005-4205-3

[197] Walsh, E.J., Schröder, T., Wallace, R.L. and Rico-Martinez, R. Cryptic speciation in *Lecane bulla* (Monogononta: Rotifera) in Chihuahuan Desert waters. Verhandlungen des Internationalen Verein Limnologie. 2009. 30 (7): 1046–1050.

[198] Guo, R. and Chen, J. 2015. Assesing the impact of dimethoate on rotifers reproduction through the pre-exposure history. Ecotoxicology and Environmental Safety. 111: 199–205. DOI: 10.1016/j.ecoenv.2014.10.023

[199] Jones, B.L., VanLoozen, J., Kim, M.H., Miles, S.J., Dunham, C.M., Williams, L.D. and Snell TW. Stress granules form in *Brachionus manjavacas* (Rotifera) in response to a variety of stressors. Comparative Biochemistry and Physiology. Part A. Molecular and Integrative Physiology. 2013. 166 (2): 375–384. DOI: 10.1016/j.cbpa.2013.07.009

[200] van der Oost, R., Beyer, J. and Vermeulen, N.P. Fish bioaccumulation and biomarkers in environmental risk assessment: a review. Environmental Toxicology and Pharmacology. 2003. 13 (2): 57–149. DOI: 10.1016/S1382-6689(02)00126-6

[201] Alvarado-Flores, J., Rico-Martínez, R., Ventura-Juárez, J., Silva-Briano, M. and Rubio-Franchini, I. Bioconcentration and localization of lead in the freshwater rotifer *Brachionus calyciflorus* Pallas 1677 (Rotifera: Monogononta). Aquatic Toxicology. 2012. 109: 127– 132.

[202] Smith, H.A. and Snell, T.W. Differential evolution of asexual and sexual females in a benign culture environment. International Review of Hydrobiology. 2014. 99 (1–2): 117–124. DOI: 10.1002/iroh.201301711

[203] Snell, T.W., Johnston, R.K., Gribble, K.E. and Mark Welch, D.B. Rotifers as experimental tools for investigating aging. Invertebrate Reproduction and Development. 2015. 59 (suppl 1): 5–10. DOI: 10.1080/07924259.2014.925516

[204] Snell, T.W. and Johnston, R.K. Glycerol extends lifespan of *Brachionus manjavacas* (Rotifera) and protects against stressors. Experimental Gerontology. 0: 47–56. DOI: 10.1016/j.exger.2014.05.005

[205] Alvarado-Flores, J., Montoya-García, M. Del R., Ventura-Juárez, J. and Rico-Martínez, R. Immunodetection of luteinizing hormone (LH), follicle stimulating hormone (FSH), thyroid stimulating hormone (TSH) and prolactin (PRL) in *Brachionus calyciflorus* (Rotifera: Monogononta). 2009. Revista de Biología Tropical. 2009. 57 (4): 1049–1058.

[206] Hanson, S.J., Stelzer, C.P., Mark Welch, D.B. and Logsdon, J.M. Comparative transcriptome analysis of obligately asexual and cyclically sexual rotifers reveals genes with putative functions in sexual reproduction, dormancy, and asexual egg production. BMC Genomics. 2013. 14: 412. DOI: 10.1186/1471-2164-14-412

[207] Smith, H.A., Burns, A.R., Shearer, T.L. and Snell, T.W. 2012. Three heat shock proteins are essential for rotifer thermotolerance. Journal of Experimental Marine Biology and Ecology. 413: 1–6

[208] Hernández-Ruiz, E., Alvarado-Flores, J., Rubio-Franchini, I., Ventura-Juárez, J. and Rico-Martínez, R. Adverse effects and bioconcentration of chromium in two freshwater rotifer species. [In preparation].

[209] Rhee, J.S., Jeong, C.B., Kim, B.M. and Lee, J.S. P-glycoprotein (P-gp) in the monogonot rotifer, *Brachionus koreanus*: molecular characterization and expression in response to pharmaceutical. Aquatic Toxicology. 2012. 114–115: 10–118. DOI: 10.1016/j.aquatox.2012.02.009

[210] Liu BR, Liou JS, Chen YJ, Huang YW, Lee HJ. Delivery of nucleic acids, proteins and nanoparticles by arginine-rich cell-penetrating peptides in rotifers. Marine Biotechnology. 2013. 15: 584–595. DOI: 10.1007/s10126-013-9509-0

[211] Jeong, C.B., Lee, J.H., Lee, J.S. and Rhee, J.S. Early expansion and expression of the lipopolysaccharide (LPS)-induced TNF-α factor (LITAF) gene family in the LPS-exposed monogonont rotifer *Brachionus koreanus*. Comparative Biochemistry and Physiology Part B Biochemistry and Molecular Biology. 2015. 188: 15–23. DOI: 10.1016/j.cbpb.2015.06.001

Chapter 5

Biomonitoring of Coral Bleaching - A Glimpse on Biomarkers for the Early Detection of Oxidative Damages in Corals

Arun Solayan

Additional information is available at the end of the chapter

http://dx.doi.org/10.5772/61831

Abstract

Corals live in a symbiotic life with single-celled algae, zooxanthelle. Anthropogenic threats and natural threat-mediated stress destabilize the photosynthetic electron transport chain resulting in an increased production of reactive oxygen species (ROS) in symbiont algae and causes coral bleaching. In this review, the early warning system and biomarkers for oxidative damages in corals are explained. The review also discusses (1) the mechanism of coral bleaching, (2) the uses of biomarkers to detect the early signs of bleaching, and (3) laboratory and field studies that are carried out on biomarkers and coral bleaching.

Keywords: Antioxidant enzymes, oxyradicals, coral bleaching, sym32gene, cytochromeP450

1. Introduction

Coral reefs are among the most productive and diverse ecosystem on earth and support myriad of fish and invertebrate species. The importance of their productivity has prompted the world conservation strategy (UNEP/WWF) to recognize coral reefs as the most essential life support system for food production, health, and other aspects of human survival and sustainable development [1,2]. Coral reefs provide a wide array of food organisms such as fish, mollusks, crabs, shrimps, and algae that are consumed by humans. The destruction of these coral reefs would definitely lead to substantial reduction in supply of animal protein in the diet of coastal population.

In general, the major hazardous threat to coral reefs can be categorized into anthropogenic and natural origin [1]. Bryant et al. [3] developed a risk index based on the impact of anthropogenic threats to the health of coral reef system, namely, coastal development and marine pollution. Under the natural threats, mortality of corals as a result of increased sea temperature is a relatively recent phenomenon that has resulted in the dramatic decline in the number of healthy reefs around the world [4]. Although various numbers of factors are proposed as a threat to coral reefs, the important toxic consequences is oxidative stress, which leads to coral bleaching [5].

The suitable way to assess sub lethal effects of oxidative stress or early detection of coral bleaching is to quantify the physiological and biochemical responses of corals as a biomarker in response to natural and anthropogenic disturbing agents. The measurement of biochemical responses (antioxidant enzymes, oxyradicals, cytochrome P450 isoforms, heat shock protein, and symbiosis-specific genes) in reefs with response to oxidative stress caused by various factors (temperature, UV radiation, and contaminants) will serve as a good biomarker for the early detection of bleaching.

2. Coral bleaching

The major reason for global degradation of corals by bleaching is a process whereby corals lose their algal symbiont or the symbionts photosynthetic pigments degrade [6]. The existence of corals is dependent on a mutualistic symbiont relationship between the individual coral polyp and a photosynthetic dinoflagellate such as zooxanthella. Zooxanthellae are intracellular residents of the tissues of coral and provide the coral with energy produced by its photosynthetic activities. In return, the coral effectively fertilizes the zooxanthella, providing nutrients in the form of ammonia and phosphates [1]. The successfully proposed model concerning a possible mechanism of coral bleaching is based on the response to oxidative stress by both components of the symbiont relationship [7]. However, understanding the structure of coral tissues could facilitate readers to know about the mechanism of coral bleaching. Corals are formed of two layers of cells known as epidermis and gastrodermis. Both these layers were covered by mucus layer and connected with porous calcium carbonate skeleton. Tissues of corals contain large populations of eukaryotic algae, bacteria, and archaea as well as numerous viruses. In the beginning of 1883, it has been reported that hard corals were associated with intracellular microscopic algae [8], and further it was identified as dinoflagellates, *Symbiodinium* [9]. *Symbiodinium* supply a large part of the energy requirements of the corals by transferring photosynthetically fixed carbon to the coral [10]. In addition, during photosynthesis, algae produce large amounts of molecular oxygen for the respiration of corals.

As mentioned in the introductory section, anthropogenic and natural threat-mediated stress can destabilize the photosynthetic electron transport chain resulting in an increased production rate of reactive oxygen species (ROS) in symbiont algae. In addition, it is worth to note that the generation of ROS occurs in the choloroplast by various mechanisms associated with an electron transfer catalyzed with photosystem I and photosystem II [11,12]. According to

Mehler reactions, hydrogen peroxide (H_2O_2) is generated by oxygen evolving complex, and these oxyradicals can easily diffuse from the algal symbiont in to the coral cytoplasm. When it happens above the threshold level, ROS will cause oxidative damage and bleaching to corals (Figure 1). Bleaching leads to high mortality and is considered as a serious threat to the health of reef ecosystem [13]. Supporting with earlier works, it is suggested that oxidative stress plays a major role in coral bleaching [1,5,14]. Although cellular-based mechanistic models concerning oxidative stress and coral bleaching are not well established, an increasing number of works have been carried out on coral symbiotic oxidative damage in relation to free radicals generated by disturbance of symbionts photosystem [5,14,15]. Hence, an effective management of the health state of coral reefs requires an early detection or biomonitoring of the oxidative stress. The suitable way to assess sublethal effects of oxidative stress or early detection of coral bleaching is to quantify the physiological and biochemical responses of corals in response to natural and anthropogenic disturbing agents.

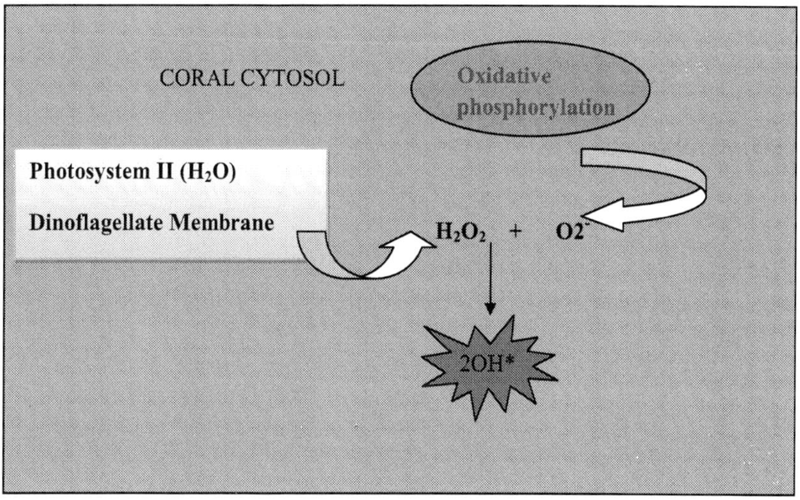

Figure 1. ROS mediated coral bleaching

3. Biomarkers

The conditions and health state of reefs are unknown since majority of them occur in remote locations [3]. It is very difficult to make a decision about the sustainable use of their resources without having an appropriate data/evidence on their health status. Hence, increased monitoring of reefs is urgently needed. We hope that biochemical responses (antioxidant enzymes, oxyradicals, fluorescent proteins, Cyp 450 isoforms, HSP, and symbiosis-specific genes) on reefs in response to oxidative stress caused by various factors (temperature, UV radiation, and contaminants) will serve as a good biomarker for the early detection of bleaching.

3.1. Oxyradicals and antioxidant enzymes

In summer, the elevation of water temperature may affect the cnidarians symbiotic life by generating oxyradicals. Ultraviolet (UV) radiation has also been shown to cause bleaching either alone or by acting synergistically with elevated temperature, wherein they produce active forms of oxygen in the zooxanthellae of corals [16]. The absorption of excitation energy in the presence of oxygen leads to the production of reactive oxygen species, ROS (O_2^-, H_2O_2), etc.

$$O_2 + e^- \rightarrow O^-_2$$
$$2O_2^- + 2H^+ \rightarrow H_2O_2 + O_2$$

ROS will further lead to the photoinhibition of photosynthesis in algae and causes bleaching in symbiotic cnidarians. Superoxide dismutase inactivates the superoxide anion by transforming it into hydrogen peroxide (H_2O_2). Hydrogen peroxide is then quickly altered by catalase and peroxidases into dioxygen (O_2) and water (H_2O). Different studies have confirmed that the production of H_2O_2 under the action of SOD is the triggering factor in the natural antioxidant defense mechanisms. SOD therefore seems to be the key enzyme in the natural defense against free radicals. Thus, antioxidant enzyme superoxide dismutase (SOD ; $2O_2^- + 2H^+ \rightarrow H_2O_2 + O_2$), catalase (CAT ; $2H_2O_2 \rightarrow 2H_2O + O_2$), glutathione peroxidase (GSH-Px ; $2GSH + ROOH \rightarrow GSSG + ROH + H_2O$), and ascorbate peroxidase are demonstrated to inactivate the oxyradicals such as O_2^- and H_2O_2 (Figure 2). In 2004, Mydlarz and Jacobs [17] revealed that H_2O_2 production occurred as an oxidative burst in a physically injured *Symbiodinium* sp. that was isolated from *Pseudopterogorgia elisabethae.* Since H_2O_2 acts as an important signaling molecule between *Symbiodinium*, i.e., zooxanthellae, and their symbiotic host, it is believed that H_2O_2 is the most important ROS associated with coral bleaching [18]. Ross et al. [19] reported that micromolar concentrations of H_2O_2 (>10 μM) induced cell death in the cyanobacterium *Microcystis aeruginosa*. In 2009, Higuchi et al. [20] studied the response of antioxidant enzymes in the coral *Galaxea fascicularis* against elevated level of H_2O_2. During short-term H_2O_2 exposure experiment, SOD and CAT activities in zooxanthellae were not significantly altered even at higher H_2O_2 concentration. Similarly, coral bleaching was not observed when exposed to H_2O_2 concentration for a period of 5 days. Both SOD and CAT activities in coral tissue and zooxanthellae were increased with high seawater temperature. It denotes that both.O_2^- and H_2O_2 were formed within the cell by the increased temperature. Further, they opined that high seawater temperature had a greater impact on the SOD and CAT activities of the coral. Anithajothi et al. [21] analyzed antioxidant enzymes (SOD, CAT, and GSH-Px) in selected scleractinian corals such as *Acropora formosa, Echinopora lamellosa, Favia favus, Favites halicora, Porites* sp., and Anacropora forbesi. They concluded that the assay of these enzymes can be used as biomarkers for identifying the susceptibility of corals towards coral bleaching. Regoli et al. [22] characterized the antioxidant efficiency of *Petrosia ficiformis* on a monthly basis by combining an analysis of the main antioxidants (superoxide dismutase, catalase, glutathione S-transferases, glutathione reductase, and glutathione peroxidases) with measurements of the total oxyradical scavenging capacity (TOSC). In summer season, significantly increased levels of catalase and TOSC were observed. The greater production of H_2O_2 in the symbioses during this period supports the hypothesis that seawater temperature can significantly modulate the pro-oxidant and antioxidant status in Mediterranean symbioses.

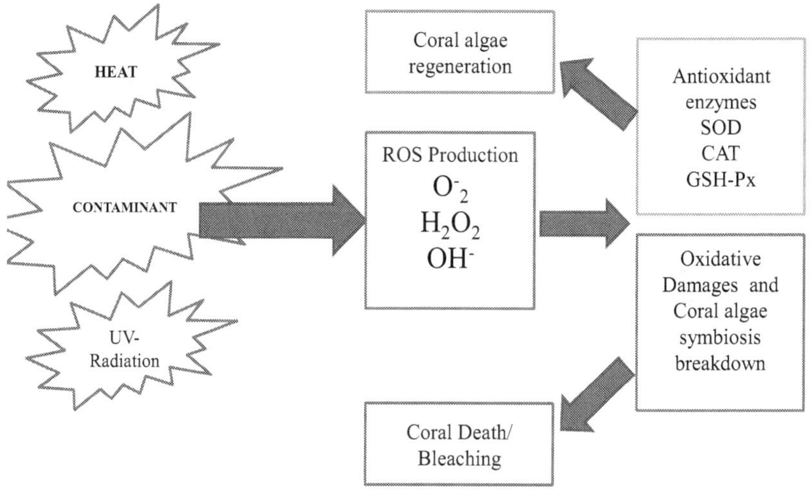

Figure 2. Oxyradicalsand antioxidant enzymes in Coral Sp.

It is very important to study the stimulation of oxyradical production in corals *in vivo* by water temperature and to what extent the oxyradicals overcomes antioxidant defenses to cause oxidative damage. No detailed study have been carried out so far on the direct measurement of oxyradicals generation *in vivo*, but an indication of such process can be obtained by detecting/analyzing the lipid peroxidation products and carbonyl proteins in heat stress exposed corals. The formation of carbonyl groups on amino acid residues as a result of free radical-initiated reactions is well documented [23]. Carbonyl formation is increased by oxidative stress and is a good marker of protein degradation [24,25]. An increased number of works have been carried out on carbonyls [26] and lipid peroxidation products [27], which shows that it could serve as a biomarker for contaminant stimulated oxidative damage. Downs et al. [7] demonstrated that heat stress causes oxidative damage in corals, which is exacerbated by exposure to light. Later in 2002 [5], they made an attempt to test whether oxidative damage is associated with coral bleaching and examined the levels of protein carbonyl and lipid peroxidation (LPO). Carbonyl protein concentration differed significantly with the exact combination of sampling date and depth and was positively correlated with ocean temperature. Lipid peroxide (LPO) also showed a similar pattern. The levels of oxidative damage products increased with water temperature and preceded coral bleaching.

3.2. Fluorescent proteins

Corals produce fluorescent proteins (FPs) that are similar to the green fluorescent protein (GFP) of jellyfish. Fluorescent protein absorbs high-energy light and protects corals. These proteins are predominantly found in scleractinian corals and constitute up to 14% of the total protein content [28]. These highly conserved molecules contain 238 amino acids that comprise 11 beta sheets and fold to form a cylinder like shape with three amino acids: serine, glycine, and tyrosine forming a posttranslationally modified fluorescent. Although the function of FPs in corals remains unclear, it is believed that it is involved in photoprotection and also acts as an antioxidant [29,30]. Blue light significantly affects corals and their symbionts. Blue light photoreceptors of corals, which are known as cryptochromes, are thought to play a role in coral bleaching during the elevation of seawater temperature. Blue light primarily damages photo system II directly and secondarily inhibits the repair of photo system II through the production of ROS [31]. The GFP of corals maximally absorbs high-energy blue light and provides photoprotection on corals. In 2009, Palmer CV and coworkers [32] found that scleractinian's fluorescent protein scavenges H_2O_2 and revealed that FPs also act as antioxidant. Carolyn Smith-Keune and Sophie Dove [33] explained that gene expression of host-specific genes such as GFP homologs may act as highly sensitive indicators for the onset of thermal stress within host coral cells. Thus, in future studies, fluorescent protein could be used as a biomarker for the early detection of thermal stress in coral reef, and based on this indication, necessary prevention steps could be taken to prevent coral bleaching.

3.3. Cytochrome P450 and monooxygenase system

Cyp 450 and flavoprotein reductase components of the microsomal mixed function oxidase (MFO) system are involves in the formation of ROS in the presence of contaminants

$$RH + O_2 + NADPH + H^+ \leftrightarrow ROH + H_2O + NADP^+.$$

It has been clearly demonstrated that algae have an ability to bioaccumulate and metabolize (via biotransformation) xenobiotic compounds through available detoxifying system such as cytochrome P 450 [34]. Also, the presence of cytochrome P 450-dependent MFO system has been documented in sea anemone and scleractinian coral [35]. CYP–carbon monoxide difference spectra have been detected for the coral species *Favia fragum, Siderastrea siderea,* and *Montastraea faveolata* [36,37]. Ramos et al. [38] analyzed the activities of cytochrome P450 and monooxygenase enzymes (CYP450, P420, and NADPH cytochrome c reductase) in corals collected from two different sampling sites (one from least contaminated site and other from contaminated site). An increased content of CYP450, P420, and NADPH cytochrome c reductase was observed in the corals collected from the contaminated site. This difference was attributed to the difference in contamination levels between the two sampling sites. Benzo(a)pyrene-induced CYP gene expression analysis in the scleractinian coral *Montastraea faveolata* [37] revealed that fuel oil exposure [39] induces CYP gene expression. Environmentally induced changes in CYP activity were observed in the coral *Stylophora pistillata* after exposure to hyposaline conditions [10] as well as in *Madracis mirabilis* after exposure to the photosynthesis inhibitor Irgarol [40]. Rosic et al. [41] discovered the presence of three new cytochrome P450 (*CYP*) genes from the reef-building coral endosymbiont *Symbiodinium*. Alteration in the expression of coral's CYP genes were analyzed during exposed to severe and moderate heat stress experiments. Samples of the scleractinian coral *Acropora millepora* were exposed to two different elevated temperatures (18-h period and 120-h period, i.e., rapid thermal stress and gradual thermal stress). The *Symbiodinium CYP* mRNA pool increased by 30% after 18 h of gradual heating and incubation at 26°C. An increase in the temperature above the average sea temperature (29°C after 72 h) resulted in a two- to fourfold increase in *CYP* expression. Both rapid thermal stress and gradual thermal stress at 32°C resulted in 50% to 90% decreases in CYP gene. The expression of CYP gene decreased under the enhanced thermal stress conditions at 32°C. These findings indicate that elevated sea temperature may affect the corals and induce the production of chemical stressors that regulate the expression of CYP genes encoding cytochrome P450 monooxygenases. This may alter the mechanism of biotransformation in corals. The studies emphasize that changes in the expression of CYP450 gene in corals could also be acted as a biomarker for the early detection of heat stress-mediated coral bleaching.

3.4. Mitochondrial integrity

Changes in environmental conditions destabilizes the symbiotic relationship between cnidarians and their dinoflagellate symbionts, *Symbiodinium* spp. As mentioned earlier, most of the studies have revealed that a breakdown in the symbiosis begins with increased ROS generation within the symbiont due to a decoupling of photosynthesis. Tchernov et al. [42] hypothesized a model for coral bleaching linking dysfunction of mitochondrial integrity to the mortality of the host animal. Mitochondria are known as batteries of the cell, which provides energy in the form of ATP and involves in ROS generation. During thermal stress, algal symbionts produce ROS that exceeds the level threshold. These molecules change the integrity of mitochondria and activate a caspase cascade within the host cell, which leads to the apoptosis and death of the corals. On the other hand, it is found that algal symbiont has the ability to remove or scavenge the ROS and gives protection from coral bleaching. It is noted that varied response was observed in the corals *Seriatopora hystrix* and *S. pistillata* to thermal stress. Although both the corals were bleached, the apoptotic response was elevated in *S. pistillata,* which resulted in the death of corals. On the contrary, apoptotic response was decreased in *S. hystrix,* which indicates that the response of corals against thermal stress is species specific and the algal symbiont of *S. hystrix* is strongly involved in scavenging ROS. However, in the case of *S. pistillata* corals, elevated ROS level induced the changes in mitochondrial integrity and further caused death (Figure 3). Dunn et al. [43] corroborate with the above said mechanism by evaluating the changes of mitochondrial integrity of host cnidarians in response to thermal stress. They assessed the overall morphology of host mitochondria associated symbionts under an experimental thermal stress using confocal and electron microscopy. It is noted that thermal stress degraded the integrity of cnidarian host mitochondria. Further, the potential sites of host mitochondrial disruption were confirmed by measuring changes in the expression of genes associated with electron transport and ATP synthesis using quantitative RT-PCR. They believed that the primary site of degradation appeared to be downstream of complex III of the electron transport chain with a significant reduction in host cytochrome *c* and ATP synthase expression. Hence, it is believed that this reduction may affect the ability of the host to remove ROS and cellular energy supplies. This finding may give us a clue on the importance of host/coral response to thermal stress and in symbiosis dysfunction that has significant implications for understanding how coral reefs will survive during the climate changes.

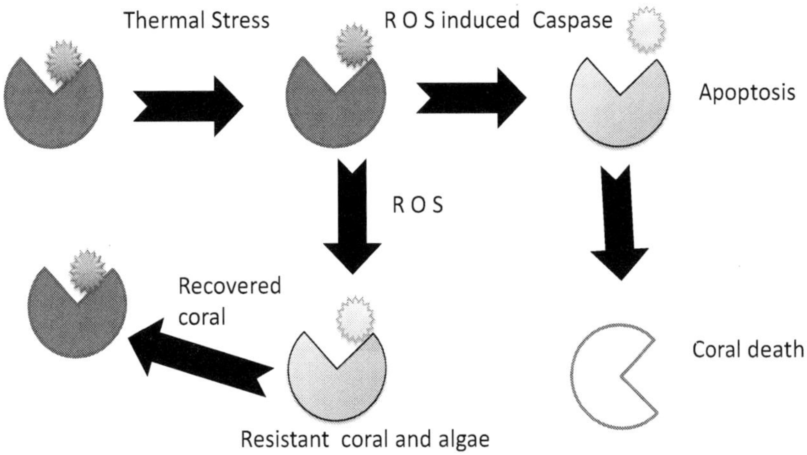

Figure 3. Thermal stress mediated coral death and recovery of ROS resistant coral Sp.

3.5. Heat Shock Proteins (HSP)

Heat stress in coral reef affects both corals and their symbionts, which further lead to bleaching of corals. Coral bleaching occurs due to the dissociation of the coral–algal symbiosis [44]. The sensitivity of coral and symbiont bond to heat stress is not well understood. However, it is believed that photosynthesis system can be impaired by heat stress [45,46]. Understanding the basic mechanism of corals against heat stress is crucial in knowing the reason of coral bleaching in response to changes in sea temperature. Heat shock protein (HSPs) represents a class of molecular chaperones that are well known for their quick response to environmental stresses [47]. Thus, alterations in coral's HSPs may serve as biological marker for heat stress. Heat shock proteins are involved in the thermotolerance of oxidative phosphorylation. Several studies demonstrate that oxidative phosphorylation is correlated with the induction of HSP. It is interesting to note that inhibitors of electron transport or inhibitors of complex I act as an inducer of HSP [48]. The mitochondrial low molecular weight HSP is usually produced only in response to environmental stress [49]. It was successfully demonstrated that chloroplast

HSP protects photosynthetic electron transport during heat stress [50], which revealed that HSPs are an important adaptation to heat stress and function as a protective molecular chaperones. Smith et al. [18] found a threefold increase in the level of HSP70 protein in host coral colony at 33°C. Chow et al. [51] also demonstrated a robust transient induction of Hsp60 in response to both light and heat stress in laminar coral. So far, the works carried out on HSP of corals provided a new insight into changes occurring in coral endosymbionts under heat stress. Further research works related to the utilization of HSP as a biomarker to thermal stress is needed.

3.6. Symbiotic-specific genes

Coral bleaching, defined as loss of color in corals, occurs due to the breakdown of the symbiosis with algae. Recently, cnidarian genes that are expressed as a function of the symbiotic state have been characterized in the sea anemone for studying cnidarian algal symbiosis [52]. They found that *sym32* gene is involved in the regulation of the symbiosis by mediating cell–cell interactions. Mitchelmore et al. [53] characterized several genes responsible for the regulation of cnidarians and their symbiotic interaction. Temperate sea anemone *Anthopleura elegantissima* has been used as a model species, and a symbiosis-specific gene, *Sym32*, was identified from the host genome. RT-PCR studies also suggested that the expression of Sym32 was correlated with the presence of host algae. No changes in algal numbers were observed on the exposure of cadmium to anemones under laboratory condition. However, they observed the downregulation of *sym 32* compared to controls. This indicates that a difference in the expression of *sym32* may act a biomarker of cnidarians–algal symbiosis breakdown.

3.7. Field and lab observations/applications of biomarkers

Corals generally grow well in clean water with a temperature between 20°C and 30°C. The optimum temperature for the growth of coral is 24°C. Coral reefs are found in great quantity in the Indian Ocean, Southeast Asia, Central Pacific, Southwest Pacific, and Caribbean regions. The largest coral reef is the Great Barrier Reef in Australia. The second largest coral reef can be found off the coast of Belize, in Central America. Coral reefs are also found in Hawaii, the Red Sea, and other areas in tropical oceans. The presence of corals in the ocean is depicted in Figure 4.

Corals and their algal endosymbionts cannot move from their habitats when they face unwanted environmental conditions such as increased seawater temperature and solar radiation. Hence, they have to develop molecular mechanisms to acclimatize and live in those unwanted conditions. Numbers of works have been carried out on coral bleaching that occurs around the world. According to the information provided by the World Resource Institute (WRI), about 370 observations were made on coral bleaching globally between 1980 and 1997. Interestingly, more than 3,700 observations were made between 1998 and 2010. This increased numbers of reports indicate the increase in awareness among researchers to monitor the health of corals and communicate about the bleaching events to the public. The suitable way to assess early detection of coral bleaching is to quantify the physiological and biochemical responses of corals as a biomarker. As mentioned in this review, changes in the biochemical parameters

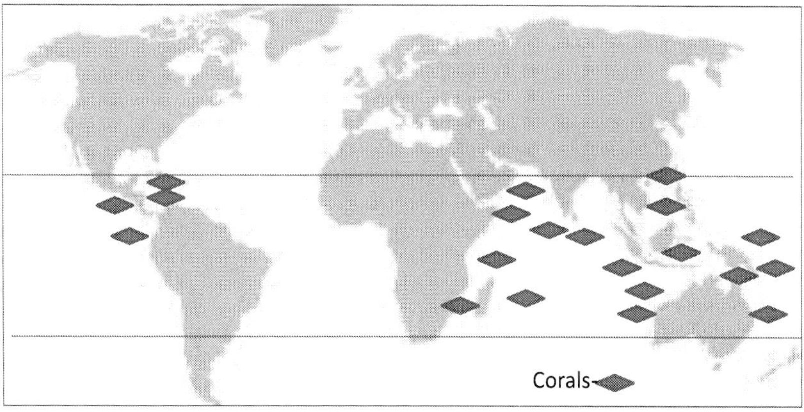

Figure 4. The presence of coral reefs in Ocean

(antioxidant enzymes, oxyradicals, cytochrome P450 isoforms, heat shock protein, and symbiosis-specific genes) of coral reefs with response to increased seawater temperature may serve as a good biomarker for the early detection of coral bleaching. Numbers of laboratory and field studies have been carried out on theses biomarkers. Some of the works relating to coral biomarkers and field applications are given in Table 1.

Location	Biomarkers	Coral host	Authors
Parque Nacional Morrocoy, Venezuela	Cytochrome-P450, Antioxidant enzymes and NADPH-C reductase	*Siderastrea siderea*	Ramos etal. (2011) [38]
Heron Island	Heat Shock Protein 70	*Porites cylindrica* and *Stylophora pistillata*	Fitt etal. (2009) [54]
Australia	Catalase	*Acropora millepora*	Krueger etal. (2015) [55]
South East Coast of India	Antioxidant enzymes	*Acropora formosa, Echinopora lamellosa, Favia favus, Favites halicora*	Anithajothi etal. (2014) [21]
France	Catalase	*Anemoniaviridis*	Merie etal. (2007) [56]
Florida	Antioxidant enzymes	Coral reef	Downs etal. (2002) [5]
USA	Fluorescence protein	*Acroporayongei*	Roth and Deheyn (2013) [57]
Great Barrier Reef	Green Fluorescence protein	Scleractinia and Alcyonacea corals	Palmer etal. (2010) [58]
Australia	Green Fluorescence protein	*Acroporamillepora*	Smith-Keune and Dove S (2008) [33]
Australia	Cytochrome P450	*Acropora millepora*	Rosic etal. (2010) [41]

Location	Biomarkers	Coral host	Authors
Honolulu, USA	Cytochrome P450	*Stylophora pistillata*	Downs etal. (2009) [10]
Italy and Maldives	Mitochondrial HSP60	*Seriatopora hystrix, Montipora monasteriata, and Acropora echinata*	Seveso etal. (2014) [59]
NJ, USA	Mitochondrial integrity	Zooxanthellate corals	Tchernov etal. (2011) [42]
USA	Heat shock protein	*Montastrea annularis*	Hayes and King (1995) [60]
Red Sea	Heat shock protein	*S. pistillata and Turbinaria reniformis*	Chow etal. (2009) [51]
USA	Heat shock protein	*Xestospongia muta*	López-Legentil etal. (2008) [61]
USA	Heat shock protein and breakdown in symbiosis between coral and zooxanthellae	*Montastraea faveolata*	DeSalvo etal. (2008) [62]
USA	Symbiosis-specific gene	*Anthopleura elegantissima*	Mitchelmore etal. (2002) [53]

Table 1. Corals and biomarkers

In the year of 2011, World Resource Institute furnished data on thermal stress affected coral reefs, which is represented in Figure 5. From the data, it can be understood that more than 40% of the corals were affected by thermal stress in Atlantic and Indian Ocean, which is higher when compared to the other regions. On viewing the earlier research works relating to biomonitoring of coral bleaching, it can be understood that only few research works were carried out in the Indian Ocean. Since corals available in this region are believed to face thermal stress, it is important to concentrate on avoiding coral bleaching in Indian Ocean. Similarly, a large volume of works has been done only on coral antioxidant enzymes and their response against climate change or thermal stress. However, an increased number of works are needed in the aspect of host symbiosis breakdown, coral's mitochondrial integrity, and cytochrome P450 protein as a biomarker of thermal stress. This may give us a better idea about coral bleaching and the utilization of biomarkers for early detection of oxidative damages. In recent days, the early prediction of thermal stress in Ocean has been proposed as the best biomarker for coral bleaching. It is very interesting to know that a computer-based model could assess sea temperature every week and predict the changes in sea temperature and warn us to take precautionary efforts to avoid temperature-mediated coral bleaching [63]. The National Oceanic and Atmospheric Administration's (NOAA) Coral Reef Watch (CRW) and the National Centers for Environment Prediction (NCEP) carried out an excellent research work to predict thermal stress that causes mass coral bleaching. In this regard, a statistical climate model to produce the first seasonal bleaching outlook system was released in 2008 at the 11th International Coral Reef Symposium. This kind of work is another milestone in this field.

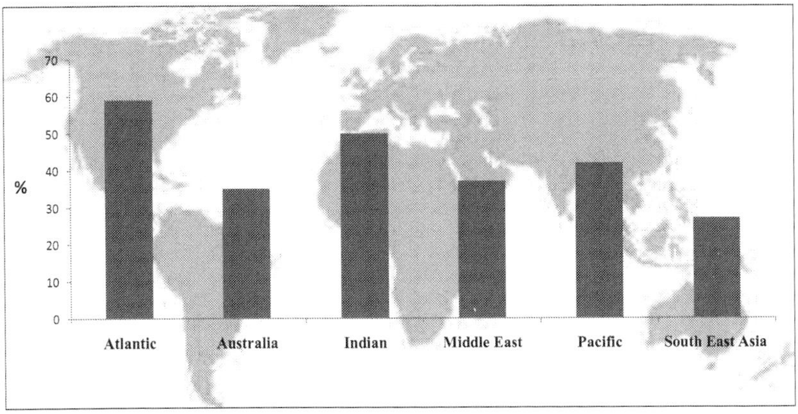

(Source: World Resource Institute (1998-2007)

Figure 5. Data on thermal stress affected coral reefs

Author details

Arun Solayan*

Address all correspondence to: sarun@rediffmail.com

Alagappa University, Karaikudi, Tamil Nadu, India

References

[1] Souter D, Obura D, Linden O. Coral Reef Degradation in the Indian Ocean: Status Report 2000. CORDIO/SAREC Marine Science, Sweden.

[2] UNEP. Co-operation for Environmental Protection in the Pacific [report]. UNEP Regional Seas Reports and Studies No. 97; 1998.

[3] Bryant D, Burke L, McManus J, Spalding M. Reefs at Risk: A Map-Based Indicator of Threats to the World's Coral Reefs. Washington, DC, World Resource Institute [WRI]; 1998; 56 p.

[4] Wilkinson CR. Global and local threats to coral reef functioning and existence: review and predictions. Marine and Freshwater Research. 1999;50(8):867–878.

[5] Downs CA, Fauth JE, Halas JC, Dustan P, Bemiss J, Woodley CM. Oxidative stress and seasonal coral bleaching. Free Radical Biology and Medicine. 2002;33:533–543.

[6] Brown, BE. Coral bleaching: causes and consequences. Coral Reefs. 1997;16(sup):129–138.

[7] Downs CA, Mueller E, Phillips S, Fauth JE, Woodley CM. A molecular biomarker system for assessing the health of coral (*Montastraea faveolata*) during heat stress. Marine Biotechnology. 2000;2:533–544.

[8] Brandt K. Über die morphologische und physiologische Bedeutung des Chlorophylls bei Tieren. Mitt Zool Stat Neapol. 1983;4:191.

[9] Klebbs, G. Ein kleiner Beitrag zur Kenntnis der Peridineen. Bot Z. 1984;10:46–47.

[10] Downs CA, Kramarsky-Winter E, Woodley CM, Downs A, Winters G, Loya Y, Ostrander GK. Cellular pathology and histopathology of hypo-salinity exposure on the coral *Stylophora pistillata*. Science of the Total Environment. 2009;407(17):4838–4851.

[11] Mehler, AH. Studies on reactions of illuminated chloroplasts. I. Mechanisms of the reduction of oxygen and other Hill reagents. Archives of Biochemistry and Biophysics. 1951;33:65–77.

[12] Richter M, Rühle W, Wild A. Studies on the mechanism of photosystem II photoinhibition II. The involvement of toxic oxygen species. Photosynthesis Research. 1990;24(3):237–243.

[13] Brown BE, Dunne RP, Goodson MS, Douglas AE. Bleaching patterns in reef corals. Nature. 2000;404(6774):142–143.

[14] Lesser MP. Oxidative stress causes coral bleaching during exposure to elevated temperatures. Coral Reefs: 1997;16:187–192.

[15] Warner ME, Fitt WK, Schnid GW. Damage to photosystem II in symbiotic dinoflagellates: a determinant of coral bleaching. Proceedings of the National Academy of Sciences of the United States of America. 1999;94(14):8007–8012.

[16] Lesser MP, Stochal WR, Tapley JDW, Shick JM. Bleaching in coral reef anthozoans: effects of irradiance, ultraviolet radiation and temperature on the activities of protective enzymes against active oxygen. Coral Reefs. 1990;8:225–232.

[17] Mydlarz LD, Jacobs RS. Comparison of an inducible oxidative burst in free-living and symbiotic dinoflagellates reveals properties of the pseudopterosins. Phytochemistry. 2004;65:3231–3241.

[18] Smith APR, Henson RNA, Rugg MID, Dolan RJ. Modulation of retrieval processing reflects accuracy of emotional source memory. Learning and Memory. 2005;12:472–479.

[19] Ross C, Santiago-Vázquez L, Paul V. Toxin release in response to oxidative stress and programmed cell death in the cyanobacterium *Microcystis aeruginosa*. Aquatic Toxicology. 2006;78(1):66–73.

[20] Higuchi T, Fujimura H, Arakaki T, Oomori T. Activities of antioxidant enzymes (SOD and CAT) in the coral *Galaxea fascicularis* against increased hydrogen peroxide concentrations in seawater. In: Proceedings of the 11th International Coral Reef Symposium. Ft. Lauderdale, Florida; 2009; pp. 926–930.

[21] Anithajothi R, Duraikannu K, Umagowsalya G, Ramakritinan CM. The presence of biomarker enzymes of selected Scleractinian corals of Palk Bay, southeast coast of India. BioMed Research International. 2014;684874:1–6.

[22] Regoli F, Cerrano C, Chierici E, Chiantore MC, Bavestrello G. Seasonal variability of prooxidant pressure and antioxidant adaptation to symbiosis in the Mediterranean demosponge *Petrosia ficiformis*. Marine Ecology Progress Series. 2004;275:129–137.

[23] Stadtman ER. Metal ion-catalyzed oxidation of proteins: biochemical mechanism and biological consequences. Free Radical Biology and Medicine. 1990;9(4):315–325.

[24] Grant AJ, Jessup W, Dean RT. Accelerated endocytosis and incomplete catabolism of radical-damaged protein. Biochimica et Biophysica Acta. 1992;1134(3):203–209.

[25] Grant AJ, Jessup W, Dean RT. Inefficient degradation of oxidized regions of protein molecules. Free Radic Res Commun. 1993;18(5):259–267

[26] Grant AJ, Jessup W, Dean RT.: Enhanced enzymatic degradation of radical damaged mitochondrial membrane components.: Free Radical Research Communications. 1993;19(2):125–134.

[27] Halliwell B, Gutteridge JMC. Free Radicals in Biology and Medicine. Oxford University Press, 1999.

[28] . Leutenegger A, D'Angelo C, Matz MV, Denzel A, Oswald F, Salih A, Nienhaus GU, Wiedenmann J. It's cheap to be colorful. Anthozoans show a slow turnover of GFP-like proteins. FEBS Journal. 2007;274(10):2496–505.

[29] Kawaguti S. On the physiology of reef corals VI. Study on the pigments. Palao Tropical Biological Station Studies. 1944;2:617–673.

[30] Salih A, Larkum A, Cox G, Kuhl M, Hoegh-Guldberg O. Fluorescent pigments in corals are photoprotective. Nature. 2000;408:850–853.

[31] Nishiyama Y, Allakhverdiev SI, Murata N. A new paradigm for the action of reactive oxygen species in the photoinhibition of photosystem II. Biochimica et Biophysica Acta. 2006;1757:742–749.

[32] Palmer CV, Roth MS, Gates RD. Red fluorescent protein responsible for pigmentation in trematode-infected *Porites compressa* tissues. Biological Bulletin. 2009;215:68.

[33] Smith-Keune C, Dove S. Gene expression of a green fluorescent protein homolog as a host-specific biomarker of heat stress within a reef-building coral. Marine Biotechnology. 2008;10(2):166–180.

[34] Kirso U, Irha N. Role of algae in fate of carcinogenic polycyclic aromatic hydrocarbons in the aquatic environment. Ecotoxicology and Environmental Safety. 1998;41(1):83–89.

[35] Heffernan LM, Winston GW. Distribution of microsomal CO-binding chromophores and EROD activity in sea anemone tissues. Marine Environmental Research. 2000;50(1–5):23–27.

[36] Garcia E, Ramos R, Bastidas C. Presence of cytochrome P450 in the Caribbean corals *Siderastrea siderea* and *Montastraea faveolata*. Ciencias Marinas. 2005;31:23–30.

[37] Ramos R, Garcia E. Induction of mixed-function oxygenase system and antioxidant enzymes in the coral *Montastraea faveolata* on acute exposure to benzo(*a*)pyrene. Comparative Biochemistry and Physiology. Part C: Toxicology and Pharmacology. 2007;144:348–355.

[38] Ramos R, Bastidas C, Debrot D, García E. *Phase I and II biotransformation and antioxidant enzymes in the coral* Siderastrea siderea *act as biomarkers for reproductive condition and habitat quality. Marine Biology Research. 2011;7(4):398–406.*

[39] Rougee L, Downs CA, Richmond RH, Ostrander GK. Alteration of normal cellular profiles in the scleractinian coral (*Pocillopora damicornis*) following laboratory exposure to fuel oil. Environmental Toxicology and Chemistry. 2006;25:3181–3187.

[40] Downs C, Downs A. Preliminary examination of short-term cellular toxicological responses of the coral *Madracis mirabilis* to acute Irgarol 1051 exposure. Archives of Environmental Contamination and Toxicology. 2007;52:47–57.

[41] Rosic NN, Pernice M, Dunn S, Dove S, Hoegh-Guldberg O. Differential regulation by heat stress of novel cytochrome P450 genes from the dinoflagellate symbionts of reef-building corals. Applied Environmental Microbiology. 2010;76(9):2823–2829.

[42] Tchernova D, Kvitta H, Haramatyd L, Thomas S, Bibbyd TS, Gorbunovd MY, Rosenfeldc H, Falkowskid PG. Apoptosis and the selective survival of host animals following thermal bleaching in zooxanthellate corals. Proceedings of the National Academy of Sciences of the United States of America. 2011;108(24):9905–9909.

[43] Dunn SR, Pernice M, Green K, Hoegh-Guldberg O, Dove SG. Thermal stress promotes host mitochondrial degradation in symbiotic cnidarians: are the batteries of the reef going to run out? PLoS One. 2012;7(7):e39024.

[44] Hoegh Guldberg O. Climate change, coral bleaching and the future of the world's coral reefs. Marine and Freshwater Research. 1999;50(8):839–866.

[45] Iglesias-Prieto R, Matta JL, Robins WA, Trench RK. The photosynthetic response to elevated temperature in the symbiotic dinoflagellate *Symbiodinium microadriaticum* in

culture. Proceedings of the National Academy of Science of USA; 1992. 89: 10302–10305.

[46] Lesser MP. Elevated temperatures and ultraviolet radiation cause oxidative stress and inhibit photosynthesis in symbiotic dinoflagellates. Limnology and Oceanography. 1996;41(2):271–283.

[47] Sørensen JG, Kristensen TN, Volker Loeschcke V. The evolutionary and ecological role of heat shock proteins. Ecology Letters. 2005;6(11):1025–1037.

[48] Ashburner M, Bonner JJ. The induction of gene activity in drosophilia by heat shock. Cell. 1979;17(2):1241–1254.

[49] Chen Q, Vierling E. Analysis of conserved domains identifies a unique structural feature of a chloroplast heat shock protein. Molecular Genetics and Genomics. 1991;226(3):425–431.

[50] Heckathorn SA, Downs CA, Sharkey TD, Coleman JS. The small, methionine-rich chloroplast heat-shock protein protects photosystem II electron transport during heat stress. Plant Physiology. 1998;116(1):439–444.

[51] Chow AM, Ferrier-Pagès C, Khalouei S, Reynaud S, Brown IR. Increased light intensity induces heat shock protein Hsp60 in coral species. Cell Stress Chaperones. 2009;14(5):469–476.

[52] Reynolds WS, Schwarz JA, Weis VM. Symbiosis-enhanced gene expression in cnidarian–algal associations: cloning and characterization of a cDNA, sym32, encoding a possible cell adhesion protein. Comparative Biochemistry and Physiology A. 2000;126(1):33–44.

[53] Mitchelmore CL, Schwarz JA, Weis VM. Development of symbiosis-specific genes as biomarkers for the early detection of cnidarian–algal symbiosis breakdown. Marine Environmental Research. 2002;54(3–5):345–349.

[54] Fitt WK, Gates RD, Hoegh-Guldberg O, Bythell JC, Jatkar A, Grottoli AG, Gomez M, Fisher P, Lajuenesse TC, Pantos O, Iglesias-Prieto R, Franklin DJ, Rodrigues LJ, Torregiani JM, Woesik RV, and Lesser MP. Response of two species of Indo-Pacific corals, *Porites cylindrica* and *Stylophora pistillata*, to short-term thermal stress: the host does matter in determining the tolerance of corals to bleaching. Journal of Experimental Marine Biology and Ecology. 373;2009:102–110.

[55] Krueger T, Hawkins TD, Becker S, Pontasch S, Dove S, Hoegh-Guldberg O, Leggat W, Fisher PL, Davy SK. Differential coral bleaching—contrasting the activity and response of enzymatic antioxidants in symbiotic partners under thermal stress. Comparative Biochemistry and Physiology A. 2015;23:190:15–25.

[56] Merle PL, Sabourault C, Richier S, Allemand D, Furla P. Catalase characterization and implication in bleaching of a symbiotic sea anemone. Free Radical Biology and Medicine. 2007;42(2):236–246.

[57] Roth MS, Deheyn DD. Effects of cold stress and heat stress on coral fluorescence in reef-building corals. Scientific Reports. 2013;3:1421.

[58] Palmer CV, Bythell JC, Willis BL. Levels of immunity parameters underpin bleaching and disease susceptibility of reef corals. FASEB Journal. 2010;24(6):1935–1946.

[59] Seveso D, Montano S, Strona G, Orlandi I, Galli P, and Vai M. The susceptibility of corals to thermal stress by analyzing Hsp60 expression. Marine Environmental Research. 2014;99:69–75.

[60] Hayes RL, King CM. Induction of 70-kD heat shock protein in scleractinian corals by elevated temperature: significance for coral bleaching. Molecular Marine Biology and Biotechnology. 1995;4(1):36–42.

[61] López-Legentil S, Song B, McMurray SE, Pawlik JR. Bleaching and stress in coral reef ecosystems: hsp70 expression by the giant barrel sponge *Xestospongia muta*. Molecular Ecology. 2008;17(7):1840–1849.

[62] DeSalvo MK, Voolstra CR, Sunagawa S, Schwarz JA, Stillman JH, Coffroth MA, Szmant AM, Medina M. Differential gene expression during thermal stress and bleaching in the Caribbean coral *Montastraea faveolata*. Molecular Ecology. 2008;17(17):3952–3971.

[63] Liu G, Matrosova LE, Penland C, Gledhill DK, Eakin CM, Webb RS, Christensen TRL, Heron SF, Morgan JA, Skirving WJ, Strong AE. NOAA Coral Reef Watch Coral Bleaching Outlook System. Proceedings of the 11th International Coral Reef Symposium, Ft. Lauderdale, Florida; July 2008; 7–11.